FRANK HARARY

Professor of Mathematics

University of Michigan

GRAPH THEORY

ADDISON-WESLEY PUBLISHING COMPANY

Reading, Massachusetts · Menlo Park, California · London · Don Mills, Ontario

This book is in the

ADDISON-WESLEY SERIES IN MATHEMATICS

Third printing, October 1972

K_5: 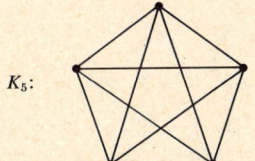 $K_{3,3}$: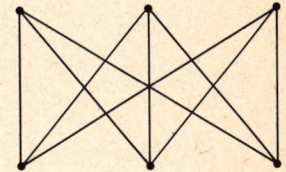

PREFACE

There are several reasons for the acceleration of interest in graph theory. It has become fashionable to mention that there are applications of graph theory to some areas of physics, chemistry, communication science, computer technology, electrical and civil engineering, architecture, operational research, genetics, psychology, sociology, economics, anthropology, and linguistics. The theory is also intimately related to many branches of mathematics, including group theory, matrix theory, numerical analysis, probability, topology, and combinatorics. The fact is that graph theory serves as a mathematical model for any system involving a binary relation. Partly because of their diagrammatic representation, graphs have an intuitive and aesthetic appeal. Although there are many results in this field of an elementary nature, there is also an abundance of problems with enough combinatorial subtlety to challenge the most sophisticated mathematician.

Earlier versions of this book have been used since 1956 when regular courses on graph theory and combinatorial theory began in the Department of Mathematics at the University of Michigan. It has been found pedagogically advantageous not to include proofs of all theorems. This device has permitted the inclusion of more theorems than would otherwise have been possible. The book can thus be used as a text in the tradition of the "Moore Method," with the student gaining mathematical power by being encouraged to prove all theorems stated without proof. Note, however, that some of the missing proofs are both difficult and long. The reader who masters the content of this book will be qualified to continue with the study of special topics and to apply graph theory to other fields.

An effort has been made to present the various topics in the theory of graphs in a logical order, to indicate the historical background, and to clarify the exposition by including figures to illustrate concepts and results. In addition, there are three appendices which provide diagrams of graphs, directed graphs, and trees. The emphasis throughout is on theorems rather than algorithms or applications, which however are occasionally mentioned.

There are vast differences in the level of exercises. Those exercises which are neither easy nor straightforward are so indicated by a bold-faced number. Exercises which are really formidable are both bold faced and starred. The reader is encouraged to consider every exercise in order to become familiar with the material which it contains. Many of the "easier" exercises may be quite difficult if the reader has not first studied the material in the chapter.

The reader is warned not to get bogged down in Chapter 2 and its many exercises, which alone can be used as a miniature course in graph theory for college freshmen or high-school seniors. The instructor can select material from this book for a one-semester course on graph theory, while the entire book can serve for a one-year course. Some of the later chapters are suitable as topics for advanced seminars. Since the elusive attribute known as "mathematical maturity" is really the only prerequisite for this book, it can be used as a text at the undergraduate or graduate level. An acquaintance with elementary group theory and matrix theory would be helpful in the last four chapters.

I owe a substantial debt to many individuals for their invaluable assistance and advice in the preparation of this book. Lowell Beineke and Gary Chartrand have been the most helpful in this respect over a period of many years! For the past year, my present doctoral students, Dennis Geller, Bennet Manvel, and Paul Stockmeyer, have been especially enthusiastic in supplying comments, suggestions, and insights. Considerable assistance was also thoughtfully contributed by Stephen Hedetniemi, Edgar Palmer, and Michael Plummer. Most recently, Branko Grünbaum and Dominic Welsh kindly gave the complete book a careful reading. I am personally responsible for all the errors and most of the off-color remarks.

Over the past two decades research support for published papers in the theory of graphs was received by the author from the Air Force Office of Scientific Research, the National Institutes of Health, the National Science Foundation, the Office of Naval Research, and the Rockefeller Foundation. During this time I have enjoyed the hospitality not only of the University of Michigan, but also of the various other scholarly organizations which I have had the opportunity to visit. These include the Institute for Advanced Study, Princeton University, the Tavistock Institute of Human Relations in London, University College London, and the London School of Economics. Reliable, rapid typing was supplied by Alice Miller and Anne Jenne of the Research Center for Group Dynamics. Finally, the author is especially grateful to the Addison-Wesley Publishing Company for its patience in waiting a full decade for this manuscript from the date the contract was signed, and for its cooperation in all aspects of the production of this book.

July 1968   F. H.

CONTENTS

I hate quotations. Tell me what you know.

R. W. Emerson

CHAPTER 1

DISCOVERY!

Eureka!
ARCHIMEDES

It is no coincidence that graph theory has been independently discovered many times, since it may quite properly be regarded as an area of applied mathematics.* Indeed, the earliest recorded mention of the subject occurs in the works of Euler, and although the original problem he was considering might be regarded as a somewhat frivolous puzzle, it did arise from the physical world. Subsequent rediscoveries of graph theory by Kirchhoff and Cayley also had their roots in the physical world. Kirchhoff's investigations of electric networks led to his development of the basic concepts and theorems concerning trees in graphs, while Cayley considered trees arising from the enumeration of organic chemical isomers. Another puzzle approach to graphs was proposed by Hamilton. After this, the celebrated Four Color Conjecture came into prominence and has been notorious ever since. In the present century, there have already been a great many rediscoveries of graph theory which we can only mention most briefly in this chronological account.

THE KÖNIGSBERG BRIDGE PROBLEM

Euler (1707–1782) became the father of graph theory as well as topology when in 1736 he settled a famous unsolved problem of his day called the Königsberg Bridge Problem. There were two islands linked to each other and to the banks of the Pregel River by seven bridges as shown in Fig. 1.1. The problem was to begin at any of the four land areas, walk across each bridge exactly once and return to the starting point. One can easily try to

---

* The basic combinatorial nature of graph theory and a clue to its wide applicability are indicated in the words of Sylvester, "The theory of ramification is one of pure colligation, for it takes no account of magnitude or position; geometrical lines are used, but have no more real bearing on the matter than those employed in genealogical tables have in explaining the laws of procreation."

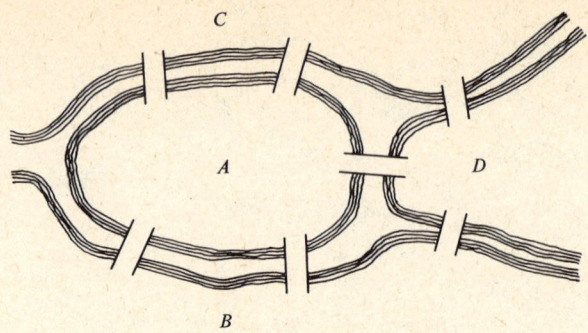

Fig. 1.1. A park in Königsberg, 1736.

solve this problem empirically, but all attempts must be unsuccessful, for the tremendous contribution of Euler in this case was negative, see [E5].

In proving that the problem is unsolvable, Euler replaced each land area by a point and each bridge by a line joining the corresponding points, thereby producing a "graph." This graph* is shown in Fig. 1.2, where the points are labeled to correspond to the four land areas of Fig. 1.1. Showing that the problem is unsolvable is equivalent to showing that the graph of Fig. 1.2 cannot be traversed in a certain way.

Rather than treating this specific situation, Euler generalized the problem and developed a criterion for a given graph to be so traversable; namely, that it is connected and every point is incident with an even number of lines. While the graph in Fig. 1.2 is connected, not every point is incident with an even number of lines.

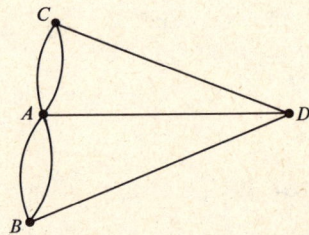

Fig. 1.2. The graph of the Königsberg Bridge Problem.

ELECTRIC NETWORKS

Kirchhoff [K7] developed the theory of trees in 1847 in order to solve the system of simultaneous linear equations which give the current in each branch and around each circuit of an electric network. Although a physicist, he thought like a mathematician when he abstracted an electric network

---

* Actually, this is a "multigraph" as we shall see in Chapter 2.

with its resistances, condensers, inductances, etc., and replaced it by its corresponding combinatorial structure consisting only of points and lines without any indication of the type of electrical element represented by individual lines. Thus, in effect, Kirchhoff replaced each electrical network by its underlying graph and showed that it is not necessary to consider every cycle in the graph of an electric network separately in order to solve the system of equations. Instead, he pointed out by a simple but powerful construction, which has since become standard procedure, that the independent cycles of a graph determined by any of its "spanning trees" will suffice. A contrived electrical network N, its underlying graph G, and a spanning tree T are shown in Fig. 1.3.

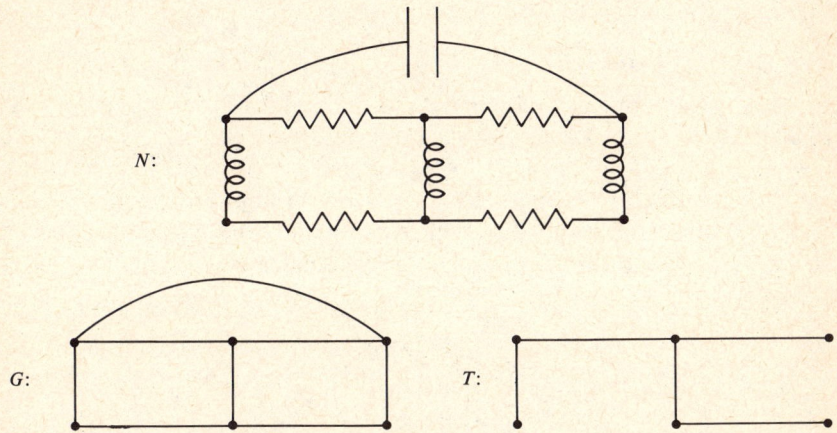

Fig. 1.3. A network N, its underlying graph G, and a spanning tree T.

CHEMICAL ISOMERS

In 1857, Cayley [C2] discovered the important class of graphs called trees by considering the changes of variables in the differential calculus. Later, he was engaged in enumerating the isomers of the saturated hydrocarbons C_nH_{2n+2}, with a given number n of carbon atoms, as shown in Fig. 1.4.

Of course, Cayley restated the problem abstractly: find the number of trees with p points in which every point has degree 1 or 4. He did not immediately succeed in solving this and so he altered the problem until he was able to enumerate: rooted trees (in which one point is distinguished from the others), trees, trees with points of degree at most 4, and finally the chemical problem of trees in which every point has degree 1 or 4, see [C3]. Jordan later (1869) independently discovered trees as a purely mathematical discipline, and Sylvester (1882) wrote that Jordan did so "without having any suspicion of its bearing on modern chemical doctrine," see [K10, p. 48].

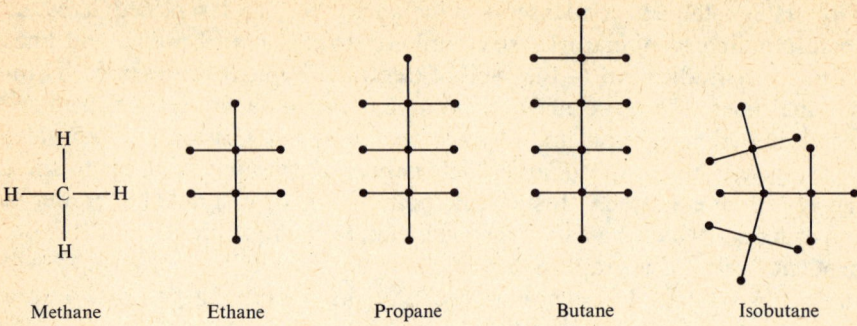

Fig. 1.4. The smallest saturated hydrocarbons.

AROUND THE WORLD

A game invented by Sir William Hamilton* in 1859 uses a regular solid dodecahedron whose 20 vertices are labeled with the names of famous cities. The player is challenged to travel "around the world" by finding a closed circuit along the edges which passes through each vertex exactly once. Hamilton sold his idea to a maker of games for 25 guineas; this was a shrewd move since the game was not a financial success.

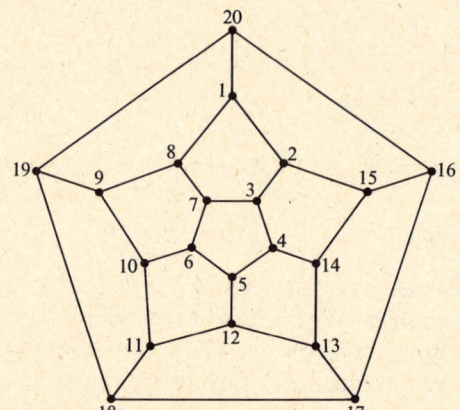

Fig. 1.5.
"Around the world."

In graphical terms, the object of the game is to find a spanning cycle in the graph of the dodecahedron, shown in Fig. 1.5. The points of the graph are marked 1, 2, ···, 20 (rather than Amsterdam, Ann Arbor, Berlin, Budapest, Dublin, Edinburgh, Jerusalem, London, Melbourne, Moscow, Novosibirsk, New York, Paris, Peking, Prague, Rio di Janeiro, Rome, San Francisco, Tokyo, and Warsaw), so that the existence of a spanning cycle is evident.

---

* See Ball and Coxeter [BC1, p. 262] for a more complete description.

THE FOUR COLOR CONJECTURE

The most famous unsolved problem in graph theory and perhaps in all of mathematics is the celebrated Four Color Conjecture. This remarkable problem can be explained in five minutes by any mathematician to the so-called man in the street. At the end of the explanation, both will understand the problem, but neither will be able to solve it.

The following quotation from the definitive historical article by May [M5] states the Four Color Conjecture and describes its role:

> [The conjecture states that] any map on a plane or the surface of a sphere can be colored with only four colors so that no two adjacent countries have the same color. Each country must consist of a single connected region, and adjacent countries are those having a boundary line (not merely a single point) in common. The conjecture has acted as a catalyst in the branch of mathematics known as combinatorial topology and is closely related to the currently fashionable field of graph theory. More than half a century of work by many (some say all) mathematicians has yielded proofs for special cases ... The consensus is that the conjecture is correct but unlikely to be proved in general. It seems destined to retain for some time the distinction of being both the simplest and most fascinating unsolved problem of mathematics.

The Four Color Conjecture has an interesting history, but its origin remains somewhat vague. There have been reports that Möbius was familiar with this problem in 1840, but it is only definite that the problem was communicated to De Morgan by Guthrie about 1850. The first of many erroneous "proofs" of the conjecture was given in 1879 by Kempe [K6]. An error was found in 1890 by Heawood [H38] who showed, however, that the conjecture becomes true when "four" is replaced by "five." A counterexample, if ever found, will necessarily be extremely large and complicated, for the conjecture was proved most recently by Ore and Stemple [OS1] for all maps with fewer than 40 countries.

The Four Color Conjecture is a problem in graph theory because every map yields a graph in which the countries (including the exterior region) are the points, and two points are joined by a line whenever the corresponding countries are adjacent. Such a graph obviously can be drawn in the plane without intersecting lines. Thus, if it is possible to color the points of every planar graph with four or fewer colors so that adjacent points have different colors, then the Four Color Conjecture will have been proved.

GRAPH THEORY IN THE 20th CENTURY

The psychologist Lewin [L2] proposed in 1936 that the "life space" of an individual be represented by a planar map.* In such a map, the regions would represent the various activities of a person, such as his work environ-

---

* Lewin used only planar maps because he always drew his figures in the plane.

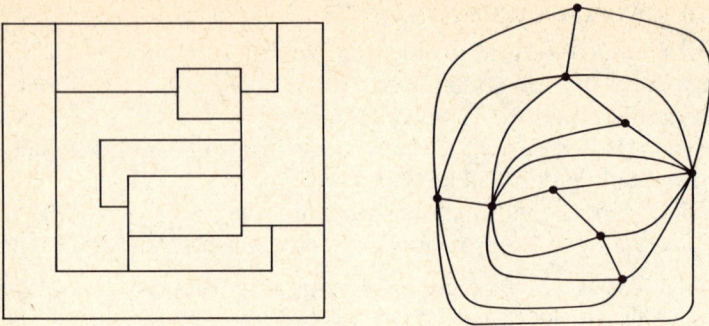

Fig. 1.6. A map and its corresponding graph.

ment, his home, and his hobbies. It was pointed out that Lewin was actually dealing with graphs, as indicated by Fig. 1.6. This viewpoint led the psychologists at the Research Center for Group Dynamics to another psychological interpretation of a graph, in which people are represented by points and interpersonal relations by lines. Such relations include love, hate, communication, and power. In fact, it was precisely this approach which led the author to a personal discovery of graph theory, aided and abetted by psychologists L. Festinger and D. Cartwright.

The world of theoretical physics discovered graph theory for its own purposes more than once. In the study of statistical mechanics by Uhlenbeck [U1], the points stand for molecules and two adjacent points indicate nearest neighbor interaction of some physical kind, for example, magnetic attraction or repulsion. In a similar interpretation by Lee and Yang [LY1], the points stand for small cubes in euclidean space, where each cube may or may not be occupied by a molecule. Then two points are adjacent whenever both spaces are occupied. Another aspect of physics employs graph theory rather as a pictorial device. Feynmann [F3] proposed the diagram in which the points represent physical particles and the lines represent paths of the particles after collisions.

The study of Markov chains in probability theory (see, for example, Feller [F2, p. 340]) involves directed graphs in the sense that events are represented by points, and a directed line from one point to another indicates a positive probability of direct succession of these two events. This is made explicit in the book [HNC1, p. 371] in which a Markov chain is defined as a network with the sum of the values of the directed lines from each point equal to 1. A similar representation of a directed graph arises in that part of numerical analysis involving matrix inversion and the calculation of eigenvalues. Examples are given by Varga [V2, p. 48]. A square matrix is given, preferably "sparse," and a directed graph is associated with it in the following way. The points denote the index of the rows and columns of the given matrix, and there is a directed line from point i to point j whenever the i, j entry of the matrix is nonzero. The similarity between this approach and that for Markov chains is immediate.

The rapidly growing fields of linear programming and operational research have also made use of a graph theoretic approach by the study of flows in networks. The books by Ford and Fulkerson [FF2], Vajda [V1] and Berge and Ghouila-Houri [BG2] involve graph theory in this way. The points of a graph indicate physical locations where certain goods may be stored or shipped, and a directed line from one place to another, together with a positive number assigned to this line, stands for a channel for the transmission of goods and a capacity giving the maximum possible quantity which can be shipped at one time.

Within pure mathematics, graph theory is studied in the pioneering book on topology by Veblen [V3, pp. 1–35]. A *simplicial complex* (or briefly a *complex*) is defined to consist of a collection V of "points" together with a prescribed collection S of nonempty subsets of V, called "simplexes," satisfying the following two conditions.

1. Every point is a simplex.
2. Every nonempty subset of a simplex is also a simplex.

The *dimension* of a simplex is one less than the number of points in it ; that of a complex is the maximum dimension of any simplex in it. In these terms, a *graph* may be defined as a complex of dimension 1 or 0. We call a 1-dimensional simplex a *line*, and note that a complex is 0-dimensional if and only if it consists of a collection of points, but no lines or other higher dimensional simplexes. Aside from these "totally disconnected" graphs, every graph is a 1-dimensional complex. It is for this reason that the subtitle of the first book ever written on graph theory [K10] is "Kombinatorische Topologie der Streckenkomplexe."

It is precisely because of the traditional use of the words point and line as undefined terms in axiom systems for geometric structures that we have chosen to use this terminology. Whenever we are speaking of "geometric" simplicial complexes as subsets of a euclidean space, as opposed to the abstract complexes defined above, we shall then use the words vertex and edge. Terminological questions will now be pursued in Chapter 2, together with some of the basic concepts and elementary theorems of graph theory.

GRAPHS

> What's in a name? That which we call a rose
> By any other name would smell as sweet.
> WILLIAM SHAKESPEARE, *Romeo and Juliet*

Most graph theorists use personalized terminology in their books, papers, and lectures. In order to avoid quibbling at conferences on graph theory, it has been found convenient to adopt the procedure that each man state in advance the graph theoretic language he would use. Even the very word "graph" has not been sacrosanct. Some authors actually define a "graph" as a graph,* but others intend such alternatives as multigraph, pseudograph, directed graph, or network. We believe that uniformity in graphical terminology will never be attained, and is not necessarily desirable.

Alas, it is necessary to present a formidable number of definitions in order to make available the basic concepts and terminology of graph theory. In addition, we give short introductions to the study of complete subgraphs, extremal graph theory (which investigates graphs with forbidden subgraphs), intersection graphs (in which the points stand for sets and nonempty intersections determine adjacency), and some useful operations on graphs.

VARIETIES OF GRAPHS

Before defining a graph, we show in Fig. 2.1 all 11 graphs with four points. Later we shall see that

i) every graph with four points is isomorphic with one of these,

ii) the 5 graphs to the left of the dashed curve in the figure are disconnected,

iii) the 6 graphs to its right are connected,

iv) the last graph is complete,

v) the first graph is totally disconnected,

vi) the first graph with four lines is a cycle,

vii) the first graph with three lines is a path.

---

* This is most frequently done by the canonical initial sentence, "In this paper we only consider finite undirected graphs without loops or multiple edges."

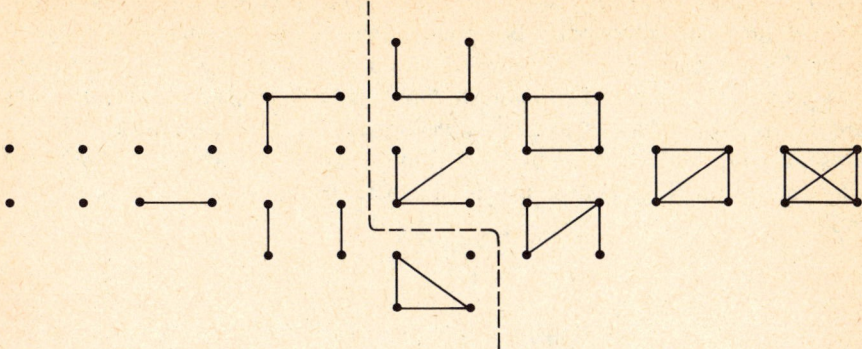

Fig. 2.1. The graphs with four points.

Rather than continue with an intuitive development of additional concepts, we proceed with the tedious but essential sequence of definition upon definition. A *graph G* consists of a finite nonempty set $V = V(G)$ of *p points** together with a prescribed set X of q unordered pairs of distinct points of V. Each pair $x = \{u, v\}$ of points in X is a *line** of G, and x is said to *join u* and v. We write $x = uv$ and say that u and v are *adjacent points* (sometimes denoted u adj v); point u and line x are *incident* with each other, as are v and x. If two distinct lines x and y are incident with a common point, then they are *adjacent lines*. A graph with p points and q lines is called a (p, q) *graph*. The $(1, 0)$ graph is *trivial*.

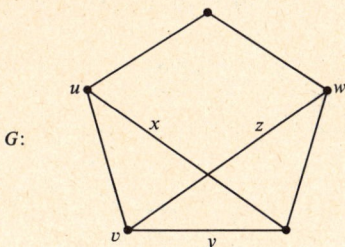

Fig. 2.2. A graph to illustrate adjacency.

It is customary to represent a graph by means of a diagram and to refer to it as the graph. Thus, in the graph G of Fig. 2.2, the points u and v are adjacent but u and w are not; lines x and y are adjacent but x and z are not. Although the lines x and z intersect in the diagram, their intersection is not a point of the graph.

---

* The following is a list of synonyms which have been used in the literature, not always with the indicated pairs:

| point, | vertex, | node, | junction, | 0-simplex, | element, |
| line, | edge, | arc, | branch, | 1-simplex, | element. |

There are several variations of graphs which deserve mention. Note that the definition of graph permits no *loop*, that is, no line joining a point to itself. In a *multigraph*, no loops are allowed but more than one line can join two points; these are called *multiple lines*. If both loops and multiple lines are permitted, we have a *pseudograph*. Figure 2.3 shows a multigraph and a pseudograph with the same "underlying graph," a triangle. We now see why the graph (Fig. 1.2) of the Königsberg bridge problem is actually a multigraph.

Fig. 2.3. A multigraph and a pseudograph.

A *directed graph* or *digraph* D consists of a finite nonempty set V of points together with a prescribed collection X of ordered pairs of distinct points. The elements of X are *directed lines* or *arcs*. By definition, a digraph has no loops or multiple arcs. An *oriented graph* is a digraph having no symmetric pair of directed lines. In Fig. 2.4 all digraphs with three points and three arcs are shown; the last two are oriented graphs. Digraphs constitute the subject of Chapter 16, but we will encounter them from time to time in the interim.

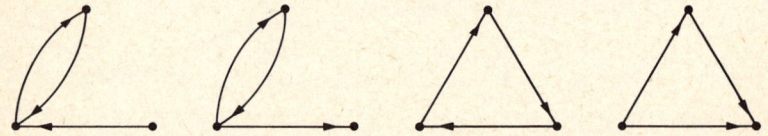

Fig. 2.4. The digraphs with three points and three arcs.

A graph G is *labeled* when the p points are distinguished from one another by names* such as $v_1, v_2, \cdots, v_p$. For example, the two graphs G_1 and G_2 of Fig. 2.5 are labeled but G_3 is not.

Two graphs G and H are *isomorphic* (written $G \cong H$ or sometimes $G = H$) if there exists a one-to-one correspondence between their point sets which preserves adjacency. For example, G_1 and G_2 of Fig. 2.5 are isomorphic under the correspondence $v_i \leftrightarrow u_i$, and incidentally G_3 is iso-

---

* This notation for points was chosen since v is the first letter of vertex. Another author calls them vertices and writes $p_1, p_2, \cdots, p_v$.

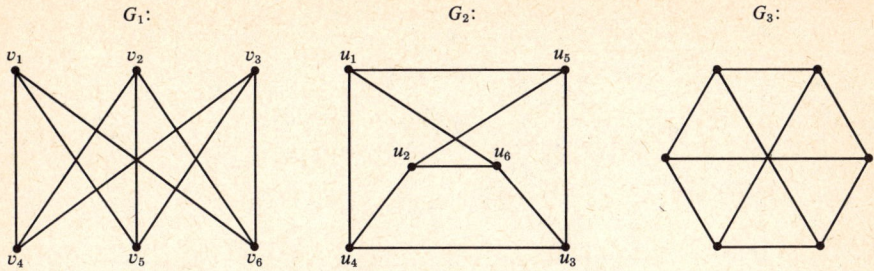

Fig. 2.5. Labeled and unlabeled graphs.

morphic with each of them. It goes without saying that isomorphism is an equivalence relation on graphs.

An *invariant* of a graph G is a number associated with G which has the same value for any graph isomorphic to G. Thus the numbers p and q are certainly invariants. A *complete set of invariants* determines a graph up to isomorphism. For example, the numbers p and q constitute such a set for all graphs with less than four points. No decent complete set of invariants for a graph is known.

A *subgraph* of G is a graph having all of its points and lines in G. If G_1 is a subgraph of G, then G is a *supergraph* of G_1. A *spanning subgraph* is a subgraph containing all the points of G. For any set S of points of G, the *induced subgraph* $\langle S \rangle$ is the maximal subgraph of G with point set S. Thus two points of S are adjacent in $\langle S \rangle$ if and only if they are adjacent in G. In Fig. 2.6, G_2 is a spanning subgraph of G but G_1 is not; G_1 is an induced subgraph but G_2 is not.

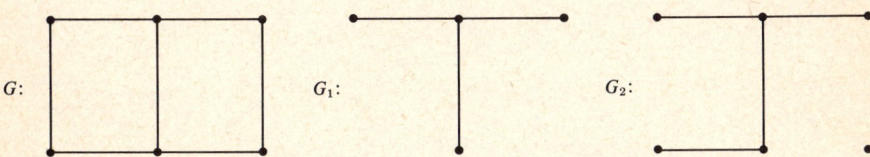

Fig. 2.6. A graph and two subgraphs.

The *removal of a point* v_i from a graph G results in that subgraph $G - v_i$ of G consisting of all points of G except v_i and all lines not incident with v_i. Thus $G - v_i$ is the maximal subgraph of G not containing v_i. On the other hand, the *removal of a line* x_j from G yields the spanning subgraph $G - x_j$ containing all lines of G except x_j. Thus $G - x_j$ is the maximal subgraph of G not containing x_j. The removal of a set of points or lines from G is defined by the removal of single elements in succession. On the other hand, if v_i and v_j are not adjacent in G, the *addition of line* v_iv_j results in the

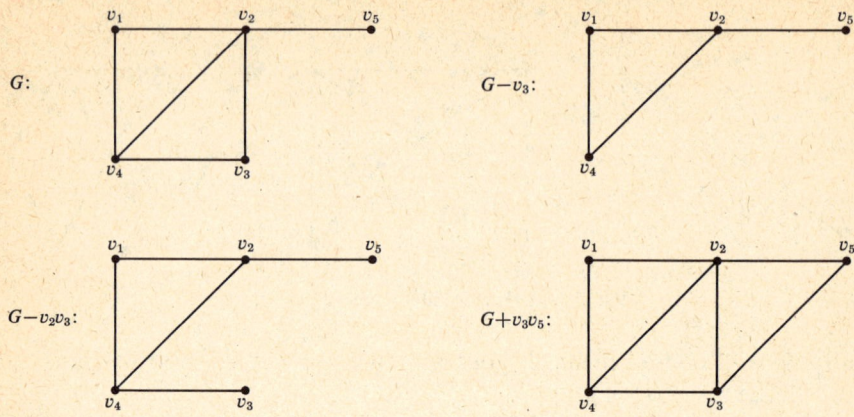

Fig. 2.7. A graph plus or minus a specific point or line.

smallest supergraph of G containing the line v_iv_j. These concepts are illustrated in Fig. 2.7.

There are certain graphs for which the result of deleting a point or line, or adding a line, is independent of the particular point or line selected. If this is so for a graph G, we denote the result accordingly by $G - v$, $G - x$, or $G + x$; see Fig. 2.8.

It was suggested by Ulam [U2, p. 29] in the following conjecture that the collection of subgraphs $G - v_i$ of G gives quite a bit of information about G itself.

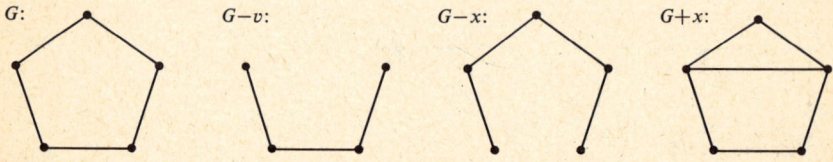

Fig. 2.8. A graph plus or minus a point or line.

Ulam's Conjecture.* Let G have p points v_i and H have p points u_i, with $p \geq 3$. If for each i, the subgraphs $G_i = G - v_i$ and $H_i = H - u_i$ are isomorphic, then the graphs G and H are isomorphic.

There is an alternative point of view to this conjecture [H29]. Draw each of the p unlabeled graphs $G - v_i$ on a 3 × 5 card. The conjecture then states that any graph from which these subgraphs can be obtained by deleting one point at a time is isomorphic to G. Thus Ulam's conjecture

---

* The reader is urged not to try to settle this conjecture since it appears to be rather difficult.

asserts that any two graphs with the same deck of cards are isomorphic. But we prefer to try to prove that from any legitimate* deck of cards, only one graph can be reconstructed.

WALKS AND CONNECTEDNESS

One of the most elementary properties that any graph can enjoy is that of being connected. In this section we develop the basic structure of connected and disconnected graphs.

A *walk* of a graph G is an alternating sequence of points and lines $v_0, x_1, v_1, \cdots, v_{n-1}, x_n, v_n$, beginning and ending with points, in which each line is incident with the two points immediately preceding and following it. This walk joins v_0 and v_n, and may also be denoted $v_0 v_1 v_2 \cdots v_n$ (the lines being evident by context); it is sometimes called a v_0–v_n walk. It is *closed* if $v_0 = v_n$ and is *open* otherwise. It is a *trail* if all the lines are distinct, and a *path* if all the points (and thus necessarily all the lines) are distinct. If the walk is closed, then it is a *cycle* provided its n points are distinct and $n \geq 3$.

In the labeled graph G of Fig. 2.9, $v_1 v_2 v_5 v_2 v_3$ is a walk which is not a trail and $v_1 v_2 v_5 v_4 v_2 v_3$ is a trail which is not a path; $v_1 v_2 v_5 v_4$ is a path and $v_2 v_4 v_5 v_2$ is a cycle.

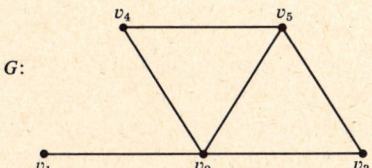

G:

Fig. 2.9. A graph to illustrate walks.

We denote by C_n the graph consisting of a cycle with n points and by P_n a path with n points; C_3 is often called a *triangle*.

A graph is *connected* if every pair of points are joined by a path. A maximal connected subgraph of G is called a *connected component* or simply a *component* of G. Thus, a disconnected graph has at least two components. The graph of Fig. 2.10 has 10 components.

The *length* of a walk $v_0 v_1 \cdots v_n$ is n, the number of occurrences of lines in it. The *girth* of a graph G, denoted $g(G)$, is the length of a shortest cycle (if any) in G; the *circumference* $c(G)$ the length of any longest cycle. Note that these terms are undefined if G has no cycles.

---

* This is a deck which can actually be obtained from some graph; another apparently difficult problem is to determine when a given deck is legitimate.

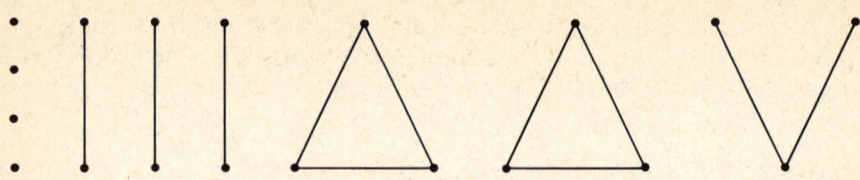

Fig. 2.10. A graph with 10 components.

The distance $d(u, v)$ between two points u and v in G is the length of a shortest path joining them if any; otherwise $d(u, v) = \infty$. In a connected graph, distance is a metric; that is, for all points u, v, and w,

1. $d(u, v) \geq 0$, with $d(u, v) = 0$ if and only if $u = v$.
2. $d(u, v) = d(v, u)$.
3. $d(u, v) + d(v, w) \geq d(u, w)$.

A shortest u–v path is often called a *geodesic*. The *diameter* $d(G)$ of a connected graph G is the length of any longest geodesic. The graph G of Fig. 2.9 has girth $g = 3$, circumference $c = 4$, and diameter $d = 2$.

The *square* G^2 of a graph G has $V(G^2) = V(G)$ with u, v adjacent in G^2 whenever $d(u, v) \leq 2$ in G. The powers G^3, G^4, $\cdots$ of G are defined similarly.

DEGREES

The *degree** of a point v_i in graph G, denoted d_i or deg v_i, is the number of lines incident with v_i. Since every line is incident with two points, it contributes 2 to the sum of the degrees of the points. We thus have a result, due to Euler [E6], which was the first theorem of graph theory!

Theorem 2.1 The sum of the degrees of the points of a graph G is twice the number of lines,

$$\sum \deg v_i = 2q. \tag{2.1}$$

Corollary 2.1(a) In any graph, the number of points of odd degree is even.†

In a (p, q) graph, $0 \leq \deg v \leq p - 1$ for every point v. The minimum degree among the points of G is denoted min deg G or $\delta(G)$ while $\Delta(G) = $ max deg G is the largest such number. If $\delta(G) = \Delta(G) = r$, then all points have the same degree and G is called *regular* of degree r. We then speak of the degree of G and write deg $G = r$.

A regular graph of degree 0 has no lines at all. If G is regular of degree 1, then every component contains exactly one line; if it is regular of degree 2,

---

* Sometimes called valency.

† The reader is reminded (see the Preface) that not all theorems are proved in the text.

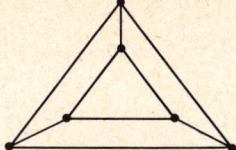 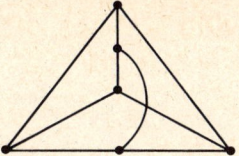

Fig. 2.11. The cubic graphs with six points.

every component is a cycle, and conversely of course. The first interesting regular graphs are those of degree 3; such graphs are called *cubic*. The two cubic graphs with six points are shown in Fig. 2.11. The second of these is isomorphic with each of the three graphs of Fig. 2.5.

Corollary 2.1(b) Every cubic graph has an even number of points.

It is convenient to have names for points of small degree. The point v is *isolated* if deg $v = 0$; it is an *endpoint* if deg $v = 1$.

THE PROBLEM OF RAMSEY

A puzzle which has become quite well known may be stated in the following form:

> Prove that at any party with six people, there are three mutual acquaintances or three mutual nonacquaintances.

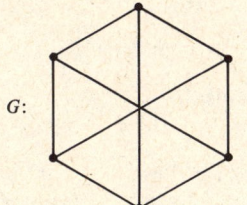 G: 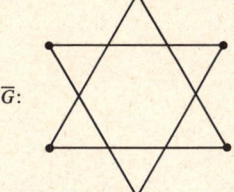 $\bar{G}$:

Fig. 2.12. A graph and its complement.

This situation may be represented by a graph G with six points standing for people, in which adjacency indicates acquaintance. Then the problem is to demonstrate that G has three mutually adjacent points or three mutually nonadjacent ones. The *complement* $\bar{G}$ of a graph G also has $V(G)$ as its point set, but two points are adjacent in $\bar{G}$ if and only if they are not adjacent in G. In Fig. 2.12, G has no triangles, while $\bar{G}$ consists of exactly two triangles.* A *self-complementary* graph is isomorphic with its complement. (See Fig. 2.13.)

---

* When drawn as $\bar{G}$ in Fig. 2.12, the union of two triangles has been called the David graph.

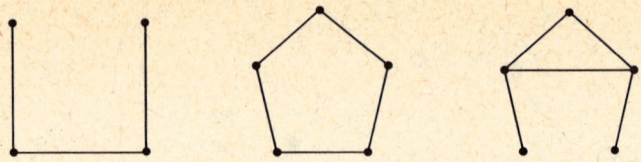

Fig. 2.13. The smallest nontrivial self-complementary graphs.

The *complete graph* K_p has every pair of its p points* adjacent. Thus K_p has $\binom{p}{2}$ lines and is regular of degree $p - 1$. As we have seen, K_3 is called a triangle. The graphs $\bar{K}_p$ are *totally disconnected*, and are regular of degree 0.

In these terms, the puzzle may be reformulated.

Theorem 2.2 For any graph G with six points, G or $\bar{G}$ contains a triangle.

Proof. Let v be a point of a graph G with six points. Since v is adjacent either in G or in $\bar{G}$ to the other five points of G, we can assume without loss of generality that there are three points u_1, u_2, u_3 adjacent to v in G. If any two of these points are adjacent, then they are two points of a triangle whose third point is v. If no two of them are adjacent in G, then u_1, u_2, and u_3 are the points of a triangle in $\bar{G}$.

The result of Theorem 2.2 suggests the general question: What is the smallest integer $r(m, n)$ such that every graph with $r(m, n)$ points contains K_m or $\bar{K}_n$?

The values $r(m, n)$ are called *Ramsey numbers*.† Of course $r(m, n) = r(n, m)$. The determination of the Ramsey numbers is an unsolved problem, although a simple bound due to Erdös and Szekeres [ES1] is known.

$$r(m, n) \leq \binom{m + n - 2}{m - 1} \qquad (2.2)$$

This problem arose from a theorem of Ramsey. An *infinite graph*‡ has an infinite point set and no loops or multiple lines. Ramsey [R2] proved (in the language of set theory) that every infinite graph contains $\aleph_0$ mutually adjacent points or $\aleph_0$ mutually nonadjacent points.

All known Ramsey numbers are given in Table 2.1, in accordance with the review article by Graver and Yakel [GY1].

---

* Since V is not empty, $p \geq 1$. Some authors admit the "empty graph" (which we would denote K_0 if it existed) and are then faced with handling its properties and specifying that certain theorems hold only for nonempty graphs, but we consider such a concept pointless.

† After Frank Ramsey, late brother of the present Archbishop of Canterbury. For a proof that $r(m, n)$ exists for all positive integers m and n, see for example Hall [H7, p. 57].

‡ Note that by definition, an infinite graph is not a graph. A review article on infinite graphs was written by Nash-Williams [N3].

Table 2.1

RAMSEY NUMBERS

$n \backslash m$	2	3	4	5	6	7
2	2	3	4	5	6	7
3	3	6	9	14	18	23
4	4	9	18			

EXTREMAL GRAPHS

The following famous theorem of Turán [T3] is the forerunner of the field of extremal graph theory, see [E3]. As usual, let $[r]$ be the greatest integer not exceeding the real number r, and $\{r\} = -[-r]$, the smallest integer not less than r.

Theorem 2.3 The maximum number of lines among all p point graphs with no triangles is $[p^2/4]$.

Proof. The statement is obvious for small values of p. An inductive proof may be given separately for odd p and for even p; we present only the latter. Suppose the statement is true for all even $p \le 2n$. We then prove it for $p = 2n + 2$. Thus, let G be a graph with $p = 2n + 2$ points and no triangles. Since G is not totally disconnected, there are adjacent points u and v. The subgraph $G' = G - \{u, v\}$ has $2n$ points and no triangles, so that by the inductive hypotheses G' has at most $[4n^2/4] = n^2$ lines. How many more lines can G have? There can be no point w such that u and v are both adjacent to w, for then u, v, and w would be points of a triangle in G. Thus if u is adjacent to k points of G', v can be adjacent to at most $2n - k$ points. Then G has at most

$$n^2 + k + (2n - k) + 1 = n^2 + 2n + 1 = p^2/4 = [p^2/4] \text{ lines.}$$

To complete the proof, we must show that for all even p, there exists a $(p, p^2/4)$ graph with no triangles. Such a graph is formed as follows: Take two sets V_1 and V_2 of $p/2$ points each and join each point of V_1 with each point of V_2. For $p = 6$, this is the graph G_1 of Fig. 2.5.

A *bigraph* (or *bipartite graph**) G is a graph whose point set V can be partitioned into two subsets V_1 and V_2 such that every line of G joins V_1 with V_2. For example, the graph of Fig. 2.14(a) can be redrawn in the form of Fig. 2.14(b) to display the fact that it is a bigraph.

If G contains every line joining V_1 and V_2, then G is a *complete bigraph*. If V_1 and V_2 have m and n points, we write $G = K_{m,n} = K(m, n)$. A *star†* is a

---

* Also called bicolorable graph, pair graph, even graph, and other things.
† When $n = 3$, Hoffman [H43] calls $K_{1,n}$ a "claw"; Erdös and Rényi [ER1], a "cherry."

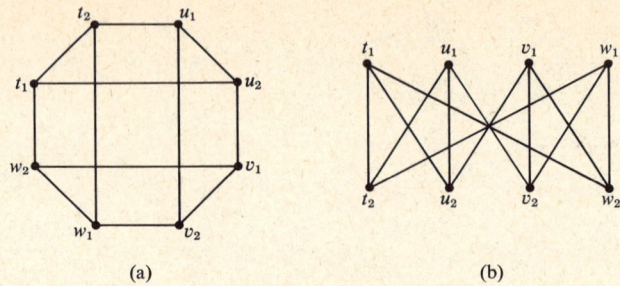

Fig. 2.14. A bigraph.

complete bigraph $K_{1,n}$. Clearly $K_{m,n}$ has mn lines. Thus, if p is even, $K(p/2, p/2)$ has $p^2/4$ lines, while if p is odd, $K([p/2], \{p/2\})$ has $[p/2]\{p/2\} = [p^2/4]$ lines. That all such graphs have no triangles follows from a theorem of König [K10, p. 170].

Theorem 2.4 A graph is bipartite if and only if all its cycles are even.

Proof. If G is a bigraph, then its point set V can be partitioned into two sets V_1 and V_2 so that every line of G joins a point of V_1 with a point of V_2. Thus every cycle $v_1 v_2 \cdots v_n v_1$ in G necessarily has its oddly subscripted points in V_1, say, and the others in V_2, so that its length n is even.

For the converse, we assume, without loss of generality, that G is connected (for otherwise we can consider the components of G separately). Take any point $v_1 \in V$, and let V_1 consist of v_1 and all points at even distance from v_1, while $V_2 = V - V_1$. Since all the cycles of G are even, every line of G joins a point of V_1 with a point of V_2. For suppose there is a line uv joining two points of V_1. Then the union of geodesics from v_1 to v and from v_1 to u together with the line uv contains an odd cycle, a contradiction.

Theorem 2.3 is the first instance of a problem in "extremal graph theory": for a given graph H, find ex (p, H), the maximum number of lines that a graph with p points can have without containing the forbidden subgraph H. Thus Theorem 2.3 states that ex $(p, K_3) = [p^2/4]$. Some other results [E3] in extremal graph theory are:

$$\text{ex } (p, C_p) = 1 + (p-1)(p-2)/2, \tag{2.3}$$

$$\text{ex } (p, K_4 - x) = [p^2/4], \tag{2.4}$$

$$\text{ex } (p, K_{1,3} + x) = [p^2/4]. \tag{2.5}$$

Turán [T3] generalized his Theorem 2.3 by determining the values of ex (p, K_n) for all $n \le p$,

$$\text{ex } (p, K_n) = \frac{(n-2)(p^2 - r^2)}{2(n-1)} + \binom{r}{2}, \tag{2.6}$$

where $p \equiv r \bmod (n - 1)$ and $0 \le r < n - 1$. A new proof of this result was given by Motzkin and Straus [MS1].

It is also known that every $(2n, n^2 + 1)$ graph contains n triangles, every $(p, 3p - 5)$ graph contains two disjoint cycles for $p \ge 6$, and every $(3n, 3n^2 + 1)$ graph contains n^2 cycles of length 4.

INTERSECTION GRAPHS

Let S be a set and $F = \{S_1, \cdots, S_p\}$ a nonempty family of distinct nonempty subsets of S whose union is S. The *intersection graph* of F is denoted $\Omega(F)$ and defined by $V(\Omega(F)) = F$, with S_i and S_j adjacent whenever $i \ne j$ and $S_i \cap S_j \ne \emptyset$. Then a graph G is *an intersection graph on S* if there exists a family F of subsets of S for which $G \cong \Omega(F)$. An early result [M4] on intersection graphs is now stated.

Theorem 2.5 Every graph is an intersection graph.

Proof. For each point v_i of G, let S_i be the union of $\{v_i\}$ with the set of lines incident with v_i. Then it is immediate that G is isomorphic with $\Omega(F)$ where $F = \{S_i\}$.

In view of this theorem, we can meaningfully define another invariant. The *intersection number* $\omega(G)$ of a given graph G is the minimum number of elements in a set S such that G is an intersection graph on S.

Corollary 2.5(a) If G is connected and $p \ge 3$, then $\omega(G) \le q$.

Proof. In this case, the points can be omitted from the sets S_i used in the proof of the theorem, so that $S = X(G)$.

Corollary 2.5(b) If G has p_0 isolated points and no K_2 components, then $\omega(G) \le q + p_0$.

The next result tells when the upper bound in Corollary 2.5(a) is attained.

Theorem 2.6 Let G be a connected graph with $p > 3$ points. Then $\omega(G) = q$ if and only if G has no triangles.

Proof. We first prove the sufficiency. In view of Corollary 2.5(a), it is only necessary to show that $\omega(G) \ge q$ for any connected G with at least 4 points having no triangles. By definition of the intersection number, G is isomorphic with an intersection graph $\Omega(F)$ on a set S with $|S| = \omega(G)$. For each point v_i of G, let S_i be the corresponding set. Because G has no triangles, no element of S can belong to more than two of the sets S_i, and $S_i \cap S_j \ne \emptyset$ if and only if v_iv_j is a line of G. Thus we can form a 1–1 correspondence between the lines of G and those elements of S which belong to exactly two sets S_i. Therefore $\omega(G) = |S| \ge q$ so that $\omega(G) = q$.

To prove the necessity, let $\omega(G) = q$ and assume that G has a triangle. Then let G_1 be a maximal triangle-free spanning subgraph of G. By the preceding paragraph, $\omega(G_1) = q_1 = |X(G_1)|$. Suppose that $G_1 = \Omega(F)$, where F is a family of subsets of some set S with cardinality q_1. Let x be a line of G not in G_1 and consider $G_2 = G_1 + x$. Since G_1 is maximal triangle-free, G_2 must have some triangle, say $u_1u_2u_3$, where $x = u_1u_3$. Denote by S_1, S_2, S_3 the subsets of S corresponding to u_1, u_2, u_3. Now if u_2 is adjacent to only u_1 and u_3 in G_1, replace S_2 by a singleton chosen from $S_1 \cap S_2$, and add that element to S_3. Otherwise, replace S_3 by the union of S_3 and any element in $S_1 \cap S_2$. In either case this gives a family F' of distinct subsets of S such that $G_2 = \Omega(F')$. Thus $\omega(G_2) \leq q_1$ while $|X(G_2)| = q_1 + 1$. If $G_2 \cong G$, there is nothing to prove. But if $G_2 \neq G$, then let

$$|X(G)| - |X(G_2)| = q_0.$$

It follows that G is an intersection graph on a set with $q_1 + q_0$ elements. However, $q_1 + q_0 = q - 1$. Thus $\omega(G) < q$, completing the proof.

The intersection number of a graph had previously been studied by Erdös, Goodman, and Pósa [EGP1]. They obtained the best possible upper bound for the intersection number of a graph with a given number of points.

Theorem 2.7 For any graph G with $p \geq 4$ points, $\omega(G) \leq [p^2/4]$.
   Their proof is essentially the same as that of Theorem 2.3.

There is an intersection graph associated with every graph which depends on its complete subgraphs. A *clique* of a graph is a maximal complete subgraph. The *clique graph* of a given graph G is the intersection graph of the family of cliques of G. For example, the graph G of Fig. 2.15 obviously has K_4 as its clique graph. However, it is not true that every graph is the clique graph of some graph, for Hamelink [H9] has shown that the same graph G is a counterexample! F. Roberts and J. Spencer have just characterized clique graphs:

Theorem 2.8 A graph G is a clique graph if and only if it contains a family F of complete subgraphs, whose union is G, such that whenever every pair of such complete graphs in some subfamily F' have a nonempty intersection, the intersection of all the members of F' is not empty.

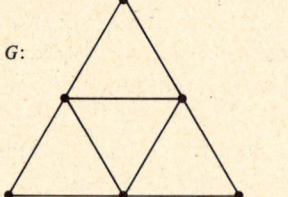
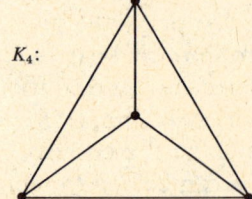

G: K_4:

Fig. 2.15.   A graph and its clique graph.

Excursion

A special class of intersection graphs was discovered in the field of genetics by Benzer [B9] when he suggested that a string of genes representing a bacterial chromosome be regarded as a closed interval on the real line. Hajós [H2] independently proposed that a graph can be associated with every finite family F of intervals S_i, which in terms of intersection graphs, is precisely $\Omega(F)$. By an *interval graph* is meant one which is isomorphic to some graph $\Omega(F)$, where F is a family of intervals. Interval graphs have been characterized by Boland and Lekkerkerker [BL2] and by Gilmore and Hoffman [GH2].

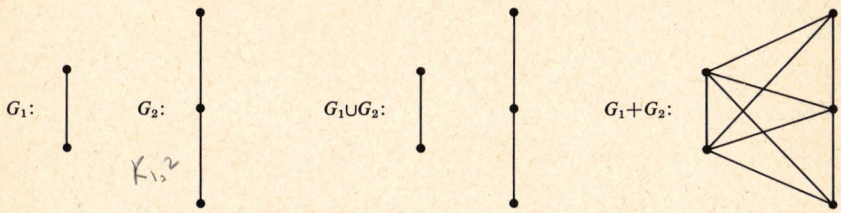

Fig. 2.16. The union and join of two graphs.

OPERATIONS ON GRAPHS

It is rather convenient to be able to express the structure of a given graph in terms of smaller and simpler graphs. It is also of value to have notational abbreviations for graphs which occur frequently. We have already introduced the complete graph K_p and its complement $\bar{K}_p$, the cycle C_n, the path P_n, and the complete bigraph $K_{m,n}$.

Throughout this section, graphs G_1 and G_2 have disjoint point sets V_1 and V_2 and line sets X_1 and X_2 respectively. Their *union** $G = G_1 \cup G_2$ has, as expected, $V = V_1 \cup V_2$ and $X = X_1 \cup X_2$. Their *join* defined by Zykov [Z1] is denoted $G_1 + G_2$ and consists of $G_1 \cup G_2$ and all lines joining V_1 with V_2. In particular, $K_{m,n} = \bar{K}_m + \bar{K}_n$. These operations are illustrated in Fig. 2.16 with $G_1 = K_2 = P_2$ and $G_2 = K_{1,2} = P_3$.

For any connected graph G, we write nG for the graph with n components each isomorphic with G. Then every graph can be written as in [HP14] in the form $\bigcup n_i G_i$ with G_i different from G_j for $i \neq j$. For example, the disconnected graph of Fig. 2.10 is $4K_1 \cup 3K_2 \cup 2K_3 \cup K_{1,2}$.

There are several operations on G_1 and G_2 which result in a graph G whose set of points is the cartesian product $V_1 \times V_2$. These include the product (or cartesian product, see Sabidussi [S5]), and the composition [H21] (or lexicographic product, see Sabidussi [S6]). Other operations† of this form are developed in Harary and Wilcox [HW1].

---

* Of course the union of two graphs which are not disjoint is also defined this way.

† These include the tensor product (Weichsel [W6], McAndrew [M7], Harary and Trauth [HT1], Brualdi [B17]), and other kinds of product defined in Berge [B12, p. 23], Ore [O5, p. 35], and Teh and Yap [TY1].

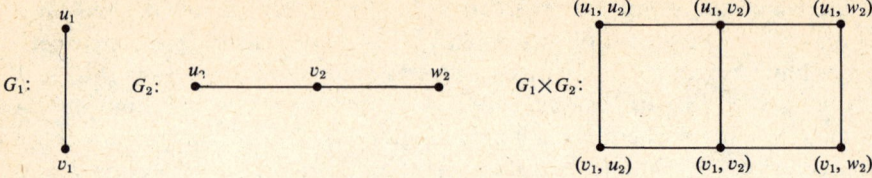

Fig. 2.17. The product of two graphs.

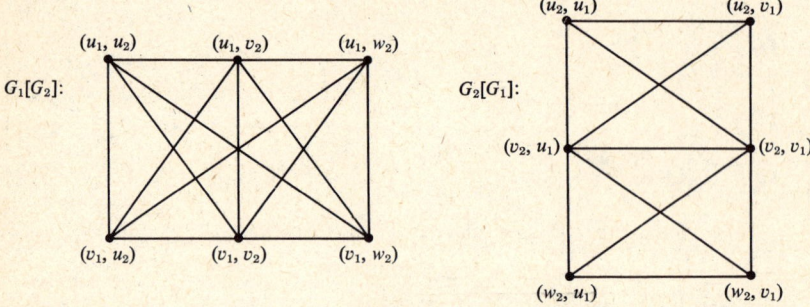

Fig. 2.18. Two compositions of graphs.

To define the *product* $G_1 \times G_2$, consider any two points $u = (u_1, u_2)$ and $v = (v_1, v_2)$ in $V = V_1 \times V_2$. Then u and v are adjacent in $G_1 \times G_2$ whenever $[u_1 = v_1$ and u_2 adj $v_2]$ or $[u_2 = v_2$ and u_1 adj $v_1]$. The product of $G_1 = P_2$ and $G_2 = P_3$ is shown in Fig. 2.17.

The *composition* $G = G_1[G_2]$ also has $V = V_1 \times V_2$ as its point set, and $u = (u_1, u_2)$ is adjacent with $v = (v_1, v_2)$ whenever $[u_1$ adj $v_1]$ or $[u_1 = v_1$ and u_2 adj $v_2]$. For the graphs G_1 and G_2 of Fig. 2.17, both compositions $G_1[G_2]$ and $G_2[G_1]$, which are obviously not isomorphic, are shown in Fig. 2.18.

If G_1 and G_2 are (p_1, q_1) and (p_2, q_2) graphs respectively, then for each of the above operations, one can calculate the number of points and lines in the resulting graph, as shown in the following table.

Table 2.2

BINARY OPERATIONS ON GRAPHS

Operation		Number of points	Number of lines
Union	$G_1 \cup G_2$	$p_1 + p_2$	$q_1 + q_2$
Join	$G_1 + G_2$	$p_1 + p_2$	$q_1 + q_2 + p_1 p_2$
Product	$G_1 \times G_2$	$p_1 p_2$	$p_1 q_2 + p_2 q_1$
Composition	$G_1[G_2]$	$p_1 p_2$	$p_1 q_2 + p_2^2 q_1$

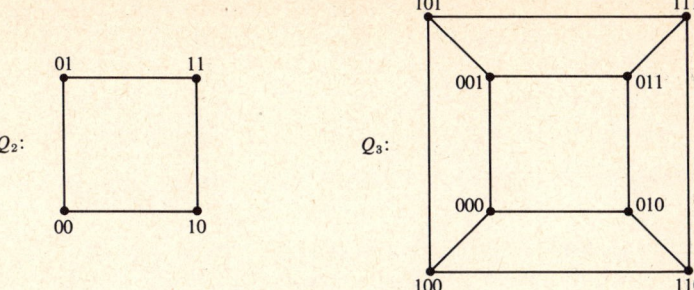

Fig. 2.19. Two cubes.

The *complete n-partite graph* $K(p_1, p_2, \cdots, p_n)$ is defined as the iterated join $\bar{K}_{p_1} + \bar{K}_{p_2} + \cdots + \bar{K}_{p_n}$. It obviously has $\Sigma\, p_i$ points and $\Sigma_{i<j}\, p_i p_j$ lines.

An especially important class of graphs known as cubes are most naturally expressed in terms of products. The *n-cube* Q_n is defined recursively by $Q_1 = K_2$ and $Q_n = K_2 \times Q_{n-1}$. Thus Q_n has 2^n points which may be labeled $a_1 a_2 \cdots a_n$, where each a_i is either 0 or 1. Two points of Q_n are adjacent if their binary representations differ at exactly one place. Figure 2.19 shows both the 2-cube and the 3-cube, appropriately labeled.

If G and H are graphs with the property that the identification of any point of G with an arbitrary point of H results in a unique graph (up to isomorphism), then we write $G \cdot H$ for this graph. For example, in Fig. 2.16 $G_2 = K_2 \cdot K_2$, while in Fig. 2.7 $G - v_3 = K_3 \cdot K_2$.

EXERCISES*

2.1 Draw all graphs with five points. (Then compare with the diagrams given in Appendix I.)

2.2 Reconstruct the graph G from its subgraphs $G_i = G - v_i$, where $G_1 = K_4 - x$, $G_2 = P_3 \cup K_1$, $G_3 = K_{1,3}$, $G_4 = G_5 = K_{1,3} + x$.

2.3 A closed walk of odd length contains a cycle.*

2.4 Prove or disprove:

   a) The union of any two distinct walks joining two points contains a cycle.
   b) The union of any two distinct paths joining two points contains a cycle.

2.5. A graph G is connected if and only if for any partition of V into two subsets V_1 and V_2, there is a line of G joining a point of V_1 with a point of V_2.

2.6 If $d(u, v) = m$ in G, what is $d(u, v)$ in the nth power G^n?

---

* Whenever a bald statement is made, it is to be proved. An exercise with number in bold face is more difficult, and one which is also starred is most difficult.

2.7 A graph H is a *square root* of G if $H^2 = G$. A graph G with p points has a square root if and only if it contains p complete subgraphs G_i such that

1. $v_i \in G_i$,
2. $v_i \in G_j$ if and only if $v_j \in G_i$,
3. each line of G is in some G_i.   (Mukhopadhyay [M18])

2.8 A finite metric space (S, d) is isomorphic to the distance space of some graph if and only if

1. The distance between any two points of S is an integer,
2. If $d(u, v) \geq 2$, then there is a third point w such that $d(u, w) + d(w, v) = d(u, v)$.   (Kay and Chartrand [KC1])

2.9 In a connected graph any two longest paths have a point in common.

2.10 It is not true that in every connected graph all longest paths have a point in common. Verify that Fig. 2.20 demonstrates this.   (Walther [W4])

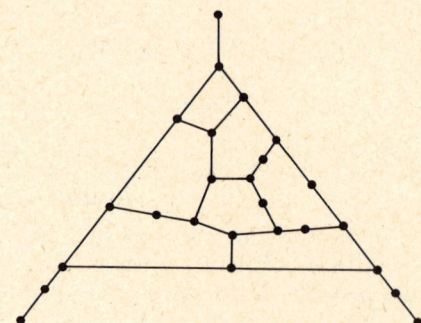

Fig. 2.20. A counterexample for Exercise 2.10.

2.11 Every graph with diameter d and girth $2d + 1$ is regular.   (Singleton [S13])

2.12 Let G be a (p, q) graph all of whose points have degree k or $k + 1$. If G has $p_k > 0$ points of degree k and p_{k+1} points of degree $k + 1$, then $p_k = (k + 1)p - 2q$.

2.13 Construct a cubic graph with $2n$ points ($n \geq 3$) having no triangles.

2.14 If G has p points and $\delta(G) \geq (p - 1)/2$, then G is connected.

2.15 If G is not connected then $\bar{G}$ is.

2.16 Every self-complementary graph has $4n$ or $4n + 1$ points.

2.17 Draw any four of the ten self-complementary graphs with eight points.

2.18 Every nontrivial self-complementary graph has diameter 2 or 3.

(Ringel [R11], Sachs [S8])

2.19 The Ramsey numbers satisfy the recurrence relation,

$$r(m, n) \leq r(m - 1, n) + r(m, n - 1).$$ (Erdös and Szekeres [ES1])

2.20 Find the maximum number of lines in a graph with p points and no even cycles.

2.21 Find the extremal graphs which do not contain K_4.   (Turán [T3])

2.22 Every $(p, p + 4)$ graph contains two line-disjoint cycles.   (Erdös [E3])

2.23 The only $(p, [p^2/4])$ graph with no triangles is $K([p/2], \{p/2\})$.

2.24 Prove or disprove: The only graph on p points with maximum intersection number is $K([p/2], \{p/2\})$.

2.25 The smallest graph having every line in at least two triangles but some line in no K_4 is the octahedron $\bar{K}_2 + C_4$.   (J. Cameron and A. R. Meetham)

2.26 Determine $\omega(K_p)$, $\omega(C_n + K_1)$, $\omega(C_n + C_n)$, and $\omega(\bar{C}_n)$.

2.27 Prove or disprove:

  a) The number of cliques of G does not exceed $\omega(G)$.
  b) The number of cliques of G is not less than $\omega(G)$.

2.28 Prove that the maximum number of cliques in a graph with p points where $p - 4 = 3r + s$, $s = 0, 1$ or 2, is $2^{2-s}3^{r+s}$.   (Moon and Moser [MM1])

2.29 A cycle of length 4 cannot be an induced subgraph of an interval graph.

2.30 Let $s(n)$ denote the maximum number of points in the n-cube which induce a cycle. Verify the following table:

n	2	3	4	5
$s(n)$	4	6	8	14

(Danzer and Klee [DK1])

2.31 Prove or disprove: If G_1 and G_2 are regular, then so is

  a) $G_1 + G_2$.   b) $G_1 \times G_2$.   c) $G_1[G_2]$.

2.32 Prove or disprove: If G_1 and G_2 are bipartite, then so is

  a) $G_1 + G_2$.   b) $G_1 \times G_2$.   c) $G_1[G_2]$.

2.33 Prove or disprove:

  a) $\overline{G_1 + G_2} = \bar{G}_1 + \bar{G}_2$.   b) $\overline{G_1 \times G_2} = \bar{G}_1 \times \bar{G}_2$.   c) $\overline{G_1[G_2]} = \bar{G}_1[\bar{G}_2]$.

2.34 a) Calculate the number of cycles in the graphs (a) $C_n + K_1$, (b) K_p, (c) $K_{m,n}$.

(Harary and Manvel [HM1])

  b) What is the maximum number of line-disjoint cycles in each of these three graphs?   (Chartrand, Geller, and Hedetniemi [CGH2])

2.35 The *conjunction* $G_1 \wedge G_2$ has $V_1 \times V_2$ as its point set and $u = (u_1, u_2)$ is adjacent to $v = (v_1, v_2)$ whenever u_1 adj v_1 and u_2 adj v_2. Then when G_1 and G_2 are connected, $G_1 \times G_2 \cong G_1 \wedge G_2$ if and only if $G_1 \cong G_2 \cong C_{2m+1}$.   (Miller [M11])

2.36 The conjunction $G_1 \wedge G_2$ of two connected graphs is connected if and only if G_1 or G_2 has an odd cycle.

*2.37 There exists a regular graph of degree r with $r^2 + 1$ points and diameter 2 only for $r = 2, 3, 7$, and possibly 57.   (Hoffman and Singleton [HS1])

*2.38 A graph G with $p = 2n$ has the property that for every set S of n points, the induced subgraphs $\langle S \rangle$ and $\langle V - S \rangle$ are isomorphic if and only if G is one of the following: K_{2n}, $K_n \times K_2$, $2K_n$, $2C_4$, and their complements.

(Kelly and Merriell [KM1])

CHAPTER 3

BLOCKS

Not merely a chip of the old block,
but the old block itself.
Edmund Burke

Some connected graphs can be disconnected by the removal of a single point, called a cutpoint. The distribution of such points is of considerable assistance in the recognition of the structure of a connected graph. Lines with the analogous cohesive property are known as bridges. The fragments of a graph held together by its cutpoints are its blocks. After characterizing these three concepts, we study two new graphs associated with a given graph: its block graph and its cutpoint graph.

CUTPOINTS, BRIDGES, AND BLOCKS

A *cutpoint* of a graph is one whose removal increases the number of components, and a *bridge* is such a line. Thus if v is a cutpoint of a connected graph G, then $G - v$ is disconnected. A *nonseparable* graph is connected, nontrivial, and has no cutpoints. A *block* of a graph is a maximal nonseparable subgraph. If G is nonseparable, then G itself is often called a block.

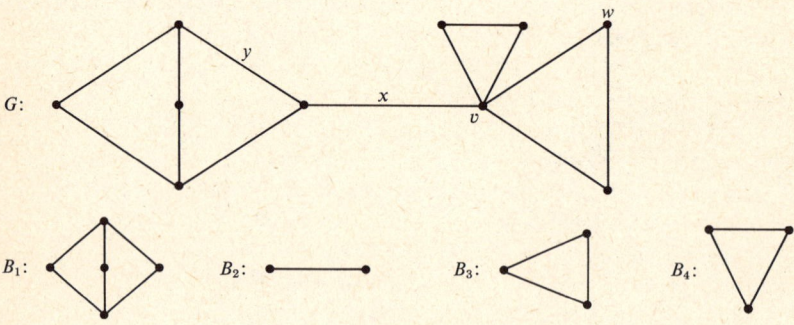

Fig. 3.1. A graph and its blocks.

In Fig. 3.1, v is a cutpoint while w is not; x is a bridge but y is not; and the four blocks of G are displayed. Each line of a graph lies in exactly one of its blocks, as does each point which is not isolated or a cutpoint. Furthermore, the lines of any cycle of G also lie entirely in a single block. Thus in particular, the blocks of a graph partition its lines and its cycles regarded as sets of lines. The first three theorems of this chapter present several equivalent conditions for each of these concepts.

Theorem 3.1 Let v be a point of a connected graph G. The following statements are equivalent:

(1) v is a cutpoint of G.

(2) There exist points u and w distinct from v such that v is on every u–w path.

(3) There exists a partition of the set of points $V - \{v\}$ into subsets U and W such that for any points $u \in U$ and $w \in W$, the point v is on every u–w path.

Proof. (*1*) *implies* (*3*) Since v is a cutpoint of G, $G - v$ is disconnected and has at least two components. Form a partition of $V - \{v\}$ by letting U consist of the points of one of these components and W the points of the others. Then any two points $u \in U$ and $w \in W$ lie in different components of $G - v$. Therefore every u–w path in G contains v.

(*3*) *implies* (*2*) This is immediate since (2) is a special case of (3).

(*2*) *implies* (*1*) If v is on every path in G joining u and w, then there cannot be a path joining these points in $G - v$. Thus $G - v$ is disconnected, so v is a cutpoint of G.

Theorem 3.2 Let x be a line of a connected graph G. The following statements are equivalent:

(1) x is a bridge of G.

(2) x is not on any cycle of G.

(3) There exist points u and v of G such that the line x is on every path joining u and v.

(4) There exists a partition of V into subsets U and W such that for any points $u \in U$ and $w \in W$, the line x is on every path joining u and w.

Theorem 3.3 Let G be a connected graph with at least three points. The following statements are equivalent:

(1) G is a block.

(2) Every two points of G lie on a common cycle.

(3) Every point and line of G lie on a common cycle.

(4) Every two lines of G lie on a common cycle.

(5) Given two points and one line of G, there is a path joining the points which contains the line.

(6) For every three distinct points of G, there is a path joining any two of them which contains the third.

(7) For every three distinct points of G, there is a path joining any two of them which does not contain the third.

Proof. (1) *implies* (2) Let u and v be distinct points of G, and let U be the set of points different from u which lie on a cycle containing u. Since G has at least three points and no cutpoints, it has no bridges; therefore, every point adjacent to u is in U, so U is not empty.

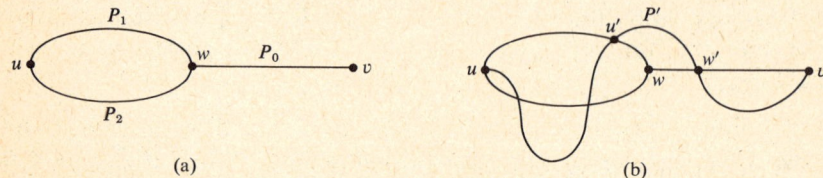

Fig. 3.2. Paths in blocks.

Suppose v is not in U. Let w be a point in U for which the distance $d(w, v)$ is minimum. Let P_0 be a shortest w–v path, and let P_1 and P_2 be the two u–w paths of a cycle containing u and w (see Fig. 3.2a). Since w is not a cutpoint, there is a u–v path P' not containing w (see Fig. 3.2b). Let w' be the point nearest u in P' which is also in P_0, and let u' be the last point of the u–w' subpath of P' in either P_1 or P_2. Without loss of generality, we assume u' is in P_1.

Let Q_1 be the u–w' path consisting of the u–u' subpath of P_1 and the u'–w' subpath of P'. Let Q_2 be the u–w' path consisting of P_2 followed by the w–w' subpath of P_0. Then Q_1 and Q_2 are disjoint u–w' paths. Together they form a cycle, so w' is in U. Since w' is on a shortest w–v path, $d(w', v) < d(w, v)$. This contradicts our choice of w, proving that u and v do lie on a cycle.

(2) *implies* (3) Let u be a point and vw a line of G. Let Z be a cycle containing u and v. A cycle Z' containing u and vw can be formed as follows. If w is on Z, then Z' consists of vw together with the v–w path of Z containing u. If w is not on Z, there is a w–u path P not containing v, since otherwise v would be a cutpoint by Theorem 3.1. Let u' be the first point of P in Z. Then Z' consists of vw followed by the w–u' subpath of P and the u'–v path in Z containing u.

(3) *implies* (4) This proof is analogous to the preceding one, and the details are omitted.

(4) implies (5) Any two points of G are incident with one line each, which lie on a cycle by (4). Hence any two points of G lie on a cycle, and we have (2), so also (3). Let u and v be distinct points and x a line of G. By statement (3), there are cycles Z_1 containing u and x, and Z_2 containing v and x. If v is on Z_1 or u is on Z_2, there is clearly a path joining u and v containing x. Thus, we need only consider the case where v is not on Z_1 and u is not on Z_2. Begin with u and proceed along Z_1 until reaching the first point w of Z_2, then take the path on Z_2 joining w and v which contains x. This walk constitutes a path joining u and v that contains x.

(5) implies (6) Let u, v, and w be distinct points of G, and let x be any line incident with w. By (5), there is a path joining u and v which contains x, and hence must contain w.

(6) implies (7) Let u, v, and w be distinct points of G. By statement (6), there is a u–w path P containing v. The u–v subpath of P does not contain w.

(7) implies (1) By statement (7), for any two points u and v, no point lies on every u–v path. Hence, G must be a block.

Theorem 3.4 Every nontrivial connected graph has at least two points which are not cutpoints.

Proof. Let u and v be points at maximum distance in G, and assume v is a cutpoint. Then there is a point w in a different component of $G - v$ than u. Hence v is in every path joining u and w, so $d(u, w) > d(u, v)$, which is impossible. Therefore v and similarly u are not cutpoints of G.

BLOCK GRAPHS AND CUTPOINT GRAPHS

There are several intersection graphs derived from a graph G which reflect its structure. If we take the blocks of G as the family F of sets, then the intersection graph $\Omega(F)$ is the *block graph* of G, denoted by $B(G)$. The blocks of G correspond to the points of $B(G)$ and two of these points are adjacent whenever the corresponding blocks contain a common cutpoint of G. On

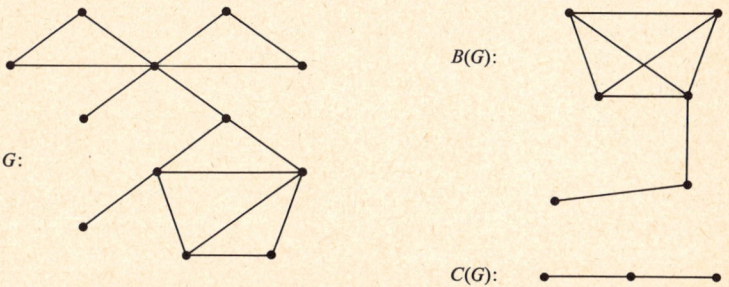

Fig. 3.3. A graph, its block graph, and its cutpoint graph.

the other hand, to obtain a graph whose points correspond to the cutpoints of G, we can take the sets S_i to be the union of all blocks which contain the cutpoint v_i. The resulting intersection graph $\Omega(F)$ is called the *cutpoint graph*, $C(G)$. Thus two points of $C(G)$ are adjacent if the cutpoints of G to which they correspond lie on a common block. Note that $C(G)$ is defined only for graphs G which have at least one cutpoint. Figure 3.3 illustrates these concepts, which were introduced in [H28].

Theorem 3.5 A graph H is the block graph of some graph if and only if every block of H is complete.

Proof. Let $H = B(G)$, and assume there is a block H_i of H which is not complete. Then there are two points in H_i which are nonadjacent and lie on a shortest common cycle Z of length at least 4. But the union of the blocks of G corresponding to the points of H_i which lie on Z is then connected and has no cutpoint, so it is itself contained in a block, contradicting the maximality property of a block of a graph.

On the other hand, let H be a given graph in which every block is complete. Form $B(H)$, and then form a new graph G by adding to each point H_i of $B(H)$ a number of endlines equal to the number of points of the block H_i which are not cutpoints of H. Then it is easy to see that $B(G)$ is isomorphic to H.

Clearly the same criterion also characterizes cutpoint graphs.

EXERCISES

3.1 What is the maximum number of cutpoints in a graph with p points?

3.2 A cubic graph has a cutpoint if and only if it has a bridge.

3.3 The smallest number of points in a cubic graph with a bridge is 10.

3.4 If v is a cutpoint of G, then v is not a cutpoint of the complement $\bar{G}$.

(Harary [H15])

3.5 A point v of G is a cutpoint if and only if there are points u and w adjacent to v such that v is on every u–w path.

3.6 Prove or disprove: A connected graph G with $p \geq 3$ is a block if and only if given any two points and one line, there is a path joining the points which does not contain the line.

3.7 A connected graph with at least two lines is a block if and only if any two adjacent lines lie on a cycle.

3.8 Let G be a connected graph with at least three points. The following statements are equivalent:

1. G has no bridges.
2. Every two points of G lie on a common closed trail.
3. Every point and line of G lie on a common closed trail.

4. Every two lines of G lie on a common closed trail.
5. For every pair of points and every line of G, there is a trail joining the points which contains the line.
6. For every pair of points and every line of G, there is a path joining the points which does not contain the line.
7. For every three points there is a trail joining any two which contains the third.

3.9 If G is a block with $\delta \geq 3$, then there is a point v such that $G - v$ is also a block.

(A. Kaugars)

3.10 The square of every nontrivial connected graph is a block.

3.11 If G is a connected graph with at least one cutpoint, then $B(B(G))$ is isomorphic to $C(G)$.

3.12 Let $b(v)$ be the number of blocks to which point v belongs in a connected graph G. Then the number of blocks of G is given by

$$b(G) - 1 = \sum [b(v) - 1]. \qquad \text{(Harary [H22])}$$

3.13 Let $c(B)$ be the number of cutpoints of a connected graph G which are points of the block B. Then the number of cutpoints of G is given by

$$c(G) - 1 = \sum [c(B) - 1]. \qquad \text{(Gallai [G3])}$$

3.14 A block G is *line-critical* if every subgraph $G - x$ is not a block. A *diagonal* of G is a line joining two points of a cycle not containing it. Let G be a line-critical block with $p \geq 4$.

a) G has no diagonals.
b) G contains no triangles.
c) $p \leq q \leq 2p - 4$.
d) The removal of all points of degree 2 results in a disconnected graph, provided G is not a cycle. (Plummer [P4])

CHAPTER 4

TREES

Poems are made by fools like me,
But only God can make a tree.

JOYCE KILMER

There is one simple and important kind of graph which has been given the same name by all authors, namely a tree. Trees are important not only for sake of their applications to many different fields, but also to graph theory itself. One reason for the latter is that the very simplicity of trees make it possible to investigate conjectures for graphs in general by first studying the situation for trees. An example is provided by Ulam's conjecture mentioned in Chapter 2.

Several ways of defining a tree are developed. Using geometric terminology, we study centrality of trees. This is followed by a discussion of a tree which is naturally associated with every connected graph: its block-cutpoint tree. Finally, we see how each spanning tree of a graph G gives rise to a collection of independent cycles of G, and mention the dual (complementary) construction of a collection of independent cocycles from each spanning cotree.

CHARACTERIZATION OF TREES

A graph is *acyclic* if it has no cycles. A *tree* is a connected acyclic graph. Any graph without cycles is a *forest*, thus the components of a forest are trees. There are 23 different trees* with eight points, as shown in Fig. 4.1. There are numerous ways of defining trees, as we shall now see.

Theorem 4.1 The following statements are equivalent for a graph G:

(1) G is a tree.

(2) Every two points of G are joined by a unique path.

---

* It is interesting to ask people to draw the trees with eight points. Some trees will frequently be missed and others duplicated.

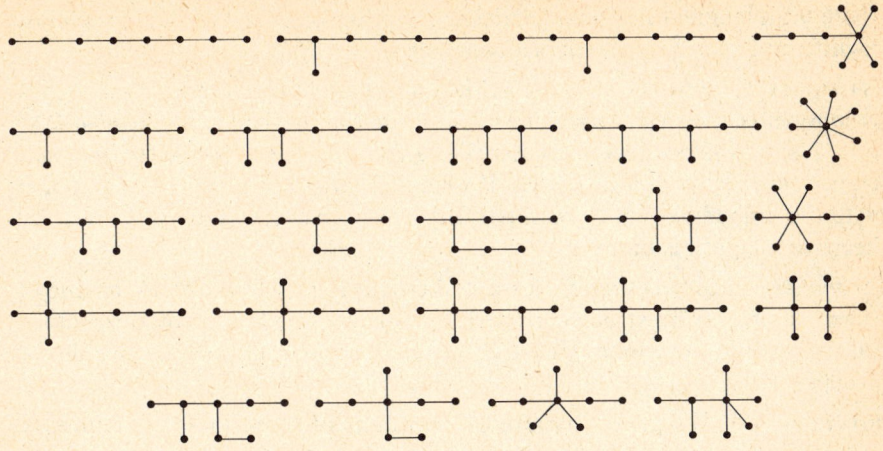

Fig. 4.1. The 23 trees with eight points.

(3) G is connected and $p = q + 1$.

(4) G is acyclic and $p = q + 1$.

(5) G is acyclic and if any two nonadjacent points of G are joined by a line x, then $G + x$ has exactly one cycle.

(6) G is connected, is not K_p for $p \geq 3$, and if any two nonadjacent points of G are joined by a line x, then $G + x$ has exactly one cycle.

(7) G is not $K_3 \cup K_1$ or $K_3 \cup K_2$, $p = q + 1$, and if any two nonadjacent points of G are joined by a line x, then $G + x$ has exactly one cycle.

Proof. (1) *implies* (2) Since G is connected, every two points of G are joined by a path. Let P_1 and P_2 be two distinct paths joining u and v in G, and let w be the first point on P_1 (as we traverse P_1 from u to v) such that w is on both P_1 and P_2 but its successor on P_1 is not on P_2. If we let w' be the next point on P_1 which is also on P_2, then the segments of P_1 and P_2 which are between w and w' together form a cycle in G. Thus if G is acyclic, there is at most one path joining any two points.

(2) *implies* (3) Clearly G is connected. We prove $p = q + 1$ by induction. It is obvious for connected graphs of one or two points. Assume it is true for graphs with fewer than p points. If G has p points, the removal of any line of G disconnects G, because of the uniqueness of paths, and in fact this new graph will have exactly two components. By the induction hypothesis each component has one more point than line. Thus the total number of lines in G must be $p - 1$.

(3) *implies* (4) Assume that G has a cycle of length n. Then there are n points and n lines on the cycle and for each of the $p - n$ points not on the cycle,

there is an incident line on a geodesic to a point of the cycle. Each such line is different, so $q \geq p$, which is a contradiction.

(4) implies (5) Since G is acyclic, each component of G is a tree. If there are k components, then, since each one has one more point than line, $p = q + k$, so $k = 1$ and G is connected. Thus G is a tree and there is exactly one path connecting any two points of G. If we add a line uv to G, that line, together with the unique path in G joining u and v, forms a cycle. The cycle is unique because the path is unique.

(5) implies (6) Since every K_p for $p \geq 3$ contains a cycle, G cannot be one of them. Graph G must be connected, for otherwise a line x could be added joining two points in different components of G, and $G + x$ would be acyclic.

(6) implies (7) We prove that every two points of G are joined by a unique path and thus, because (2) implies (3), $p = q + 1$. Certainly every two points of G are joined by some path. If two points of G are joined by two paths, then by the proof that (1) implies (2), G has a cycle. This cycle cannot have four or more points because, if it did, then we could produce more than one cycle in $G + x$ by taking x joining two nonadjacent points on the cycle (if there are no nonadjacent points on the cycle, then G itself has more than one cycle). So the cycle is K_3, which must be a proper subgraph of G since by hypothesis G is not complete with $p \geq 3$. Since G is connected, we may assume there is another point in G which is joined to a point of this K_3. Then it is clear that if any line can be added to G, then one may be added so as to form at least two cycles in $G + x$. If no more lines may be added, so that the second condition on G is trivially satisfied, then G is K_p with $p \geq 3$, contrary to hypothesis.

(7) implies (1) If G has a cycle, that cycle must be a triangle which is a component of G, by an argument in the preceding paragraph. This component has three points and three lines. All other components of G must be trees and, in order to make $p = q + 1$, there can be only one other component. If this tree contains a path of length 2, it will be possible to add a line x to G and obtain two cycles in $G + x$. Thus this tree must be either K_1 or K_2. So G must be $K_3 \cup K_1$ or $K_3 \cup K_2$, which are the graphs which have been excluded. Thus G is acyclic. But if G is acyclic and $p = q + 1$, then G is connected since (4) implies (5) implies (6). So G is a tree, and the theorem is proved.

Because a nontrivial tree has $\Sigma\, d_i = 2q = 2(p - 1)$, there are at least two points with degree less than 2.

Corollary 4.1(a) Every nontrivial tree has at least two endpoints.

This result also follows from Theorem 3.4.

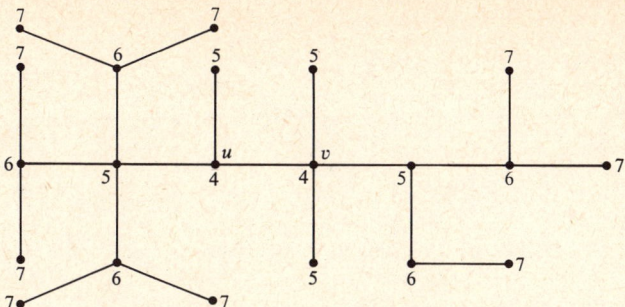

Fig. 4.2. The eccentricities of the points of a tree.

CENTERS AND CENTROIDS

The *eccentricity* $e(v)$ of a point v in a connected graph G is max $d(u, v)$ for all u in G. The *radius* $r(G)$ is the minimum eccentricity of the points. Note that the maximum eccentricity is the diameter. A point v is a *central point* if $e(v) = r(G)$, and the *center* of G is the set of all central points.

In the tree of Fig. 4.2, the eccentricity of each point is shown. This tree has diameter 7, radius 4, and the center consists of the two points u and v, each with minimum eccentricity 4. The fact that u and v are adjacent illustrates a result discovered by Jordan* and independently by Sylvester; see König [K10, p. 64].

Theorem 4.2 Every tree has a center consisting of either one point or two adjacent points.

Proof. The result is obvious for the trees K_1 and K_2. We show that any other tree T has the same central points as the tree T' obtained by removing all endpoints of T. Clearly, the maximum of the distances from a given point u of T to any other point v of T will occur only when v is an endpoint.

Thus, the eccentricity of each point in T' will be exactly one less than the eccentricity of the same point in T. Hence the points of T which possess minimum eccentricity in T are the same points having minimum eccentricity in T', that is, T and T' have the same center. If the process of removing endpoints is repeated, we obtain successive trees having the same center as T. Since T is finite, we eventually obtain a tree which is either K_1 or K_2. In either case all points of this ultimate tree constitute the center of T which thus consists of just a single point or of two adjacent points.

A *branch at a point* u of a tree T is a maximal subtree containing u as an endpoint. Thus the number of branches at u is deg u. The *weight at a point* u of T is the maximum number of lines in any branch at u. The weights at the

---

* Of Jordan Curve Theorem fame.

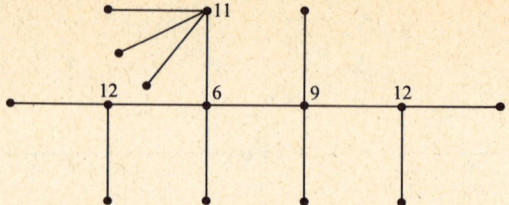

Fig. 4.3. The weights at the points of a tree.

nonendpoints of the tree in Fig. 4.3 are indicated. Of course the weight at each endpoint is 14, the number of lines.

A point v is a *centroid point* of a tree T if v has minimum weight, and the *centroid* of T consists of all such points. Jordan [J2] also proved a theorem on the centroid of a tree analogous to his result for centers.

Theorem 4.3 Every tree has a centroid consisting of either one point or two adjacent points.

The smallest trees with one and two central and centroid points are shown in Fig. 4.4.

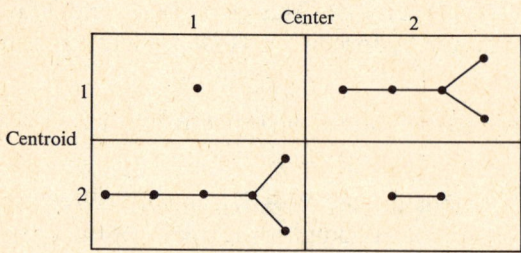

Fig. 4.4. Trees with all combinations of one or two central and centroid points.

BLOCK-CUTPOINT TREES

It has often been observed that a connected graph with many cutpoints bears a resemblance to a tree. This idea can be made more definite by associating with every connected graph a tree which displays the resemblance.

For a connected graph G with blocks $\{B_i\}$ and cutpoints $\{c_j\}$, the *block-cutpoint graph* of G, denoted by $bc(G)$, is defined as the graph having point set $\{B_i\} \cup \{c_j\}$, with two points adjacent if one corresponds to a block B_i and the other to a cutpoint c_j and c_j is in B_i. Thus $bc(G)$ is a bigraph. This concept was introduced in Harary and Prins [HP22] and also in Gallai [G3]. (See Fig. 4.5.)

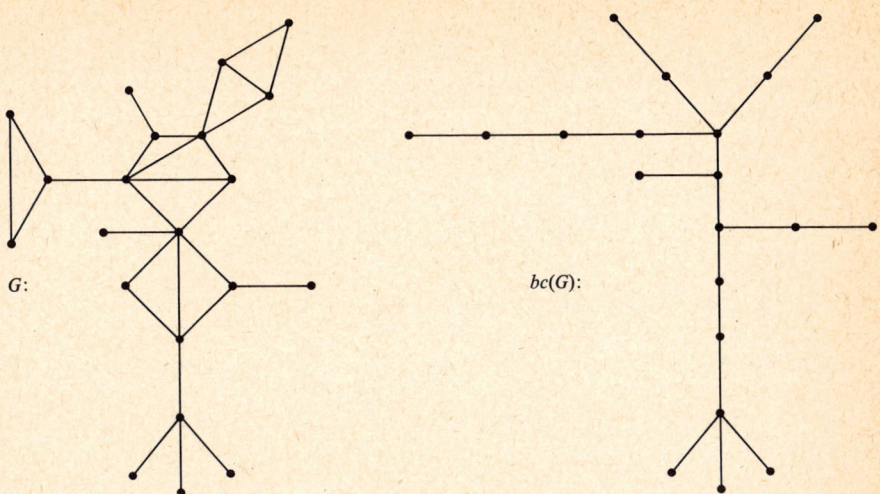

Fig. 4.5. A graph and its block-cutpoint graph.

Theorem 4.4 A graph G is the block-cutpoint graph of some graph H if and only if it is a tree in which the distance between any two endpoints is even.

In view of this theorem, we will speak of the *block-cutpoint tree* of a graph.

INDEPENDENT CYCLES AND COCYCLES

We describe two vector spaces associated with a graph G: its "cycle space" and "cocycle space." For convenience, these two vector spaces will be taken over the two element field $F_2 = \{0, 1\}$, in which $1 + 1 = 0$ (even though the theory can be modified to hold for an arbitrary field). In particular, the ε_i which occur repeatedly in the following definitions are always either 0 or 1.

As usual, let G be a graph with points $v_1, \cdots, v_p$ and lines $x_1, \cdots, x_q$. A *0-chain* of G is a formal linear combination $\Sigma \, \varepsilon_i v_i$ of points and a *1-chain* is a sum $\Sigma \, \varepsilon_i x_i$ of lines. The *boundary operator* ∂ sends 1-chains to 0-chains according to the rules:

a) ∂ is linear.

b) if $x = uv$, then $\partial x = u + v$.

On the other hand, the *coboundary operator* δ sends 0-chains to 1-chains by the rules:

a) δ is linear.

b) $\delta v = \Sigma \, \varepsilon_i x_i$, where $\varepsilon_i = 1$ whenever x_i is incident with v.

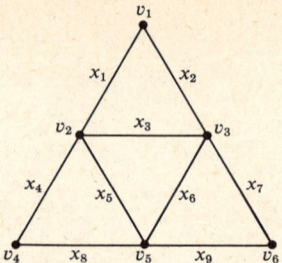

Fig. 4.6. A graph to illustrate the boundary and coboundary operators.

In Fig. 4.6, the 1-chain $\sigma_1 = x_1 + x_2 + x_4 + x_9$ has boundary

$$\partial\sigma_1 = (v_1 + v_2) + (v_1 + v_3) + (v_2 + v_4) + (v_5 + v_6)$$
$$= v_3 + v_4 + v_5 + v_6,$$

and the 0-chain $\sigma_0 = v_3 + v_4 + v_5 + v_6$ has as its coboundary

$$\delta\sigma_0 = (x_2 + x_3 + x_6 + x_7) + (x_4 + x_8)$$
$$+ (x_5 + x_6 + x_8 + x_9) + (x_7 + x_9)$$
$$= x_2 + x_3 + x_4 + x_5.$$

A 1-chain with boundary 0 is a *cycle vector** of G and can be regarded as a set of line-disjoint cycles. The collection of all cycle vectors forms a vector space over F_2 called the *cycle space* of G. A *cycle basis* of G is defined as a basis for the cycle space of G which consists entirely of cycles. We say a cycle-vector Z depends on the cycles $Z_1, Z_2, \cdots, Z_k$ if it can be written as $\sum_{i=1}^{k} \varepsilon_i Z_i$. Thus a cycle basis of G is a maximal collection of independent cycles of G, or a minimal collection of cycles on which all cycles depend.

A *cutset* of a connected graph is a collection of lines whose removal results in a disconnected graph. A *cocycle* is a minimal cutset. A *coboundary* of G is the coboundary of some 0-chain in G. The coboundary of a collection U of points is just the set of all lines joining a point in U to a point not in U. Thus every coboundary is a cutset. Since we define a cocycle as a minimal cutset of G and any minimal cutset is a coboundary, we see that a cocycle is just a minimal nonzero coboundary. The collection of all coboundaries of G is called the *cocycle space* of G, and a basis for this space which consists entirely of cocycles is called a *cocycle basis* for G.

We proceed to construct for the cycle space of G a basis which corresponds to a spanning tree T. In a connected graph G, a *chord* of a spanning tree T is a line of G which is not in T. Clearly the subgraph of G consisting of T and any chord of T has exactly one cycle. Moreover, the set $Z(T)$

---

* Most topologists and some graph theorists call this a "cycle." They then use "circuits" or "elementary cycles" or "polygons" for our cycles.

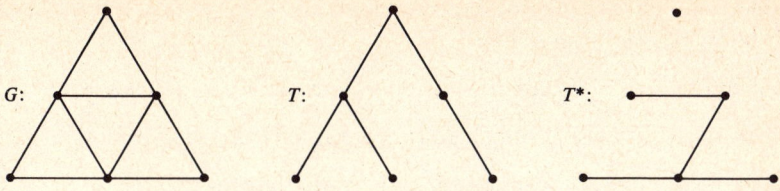

Fig. 4.7.   Graph, tree, and cotree.

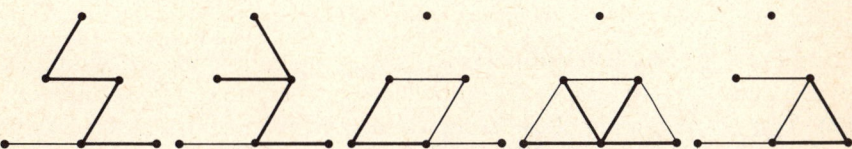

Fig. 4.8.   A cocycle basis for G of Fig. 4.7.

of cycles obtained in this way (one from each chord) is independent, since each contains a line not in any of the others. Also, every cycle Z depends on the set $Z(T)$, for Z is the symmetric difference of the cycles determined by the chords of T which lie in Z. Thus if we define $m(G)$, the *cycle rank*, to be the number of cycles in a basis for the cycle space of G, we have the following result.

Theorem 4.5   The cycle rank of a connected graph G is equal to the number of chords of any spanning tree in G.

Corollary 4.5(a)   If G is a connected (p, q) graph, then $m(G) = q - p + 1$.

Corollary 4.5(b)   If G is a (p, q) graph with k components, then

$$m(G) = q - p + k.$$

Similar results are true for the cocycle space. The *cotree* T^* of a spanning tree T in a connected graph G is the spanning subgraph of G containing exactly those lines of G which are not in T. A cotree of G is the cotree of some spanning tree T. In Fig. 4.7, a spanning tree T and its cotree T^* are displayed for the same graph G as in Fig. 4.6. The lines of G which are not in T^* are called its *twigs*. The subgraph of G consisting of T^* and any one of its twigs contains exactly one cocycle. The collection of cocycles obtained by adding twigs to T^*, one at a time, is seen to be a basis for the cocycle space of G. This is illustrated in Fig. 4.8 for the graph G and cotree T^* of Fig. 4.7, with the cocycles indicated by heavy lines. The *cocycle rank* $m^*(G)$ is the number of cocycles in a basis for the cocycle space of G.

Theorem 4.6 The cocycle rank of a connected graph G is the number of twigs in any spanning tree of T.

As in the case of cycles, we have two immediate corollaries.

Corollary 4.6(a)   If G is a connected (p, q) graph, then $m^*(G) = p - 1$.

Corollary 4.6(b)   If G is a (p, q) graph with k components, then $m^*(G) = p - k$.

Excursion

The 1-dimensional case of an important general result about simplicial complexes can be derived from Theorem 4.5. The Euler-Poincaré equation

$$\alpha_0 - \alpha_1 + \alpha_2 - \cdots = \beta_0 - \beta_1 + \beta_2 - \cdots,$$

where the β_n are the Betti numbers and the α_n are the numbers of simplexes of each dimension, holds for every simplicial complex. Recall from Chapter 1 that every graph is a simplicial complex, with its points 0-simplexes and its lines 1-simplexes. For a graph, $\beta_0 = k$, the number of connected components, and $\beta_1 = m(G)$, the number of independent cycles of G. Since no graph contains an n-simplex with $n > 1$, $\alpha_n = \beta_n = 0$, for all $n > 1$. Thus $\alpha_0 - \alpha_1 = \beta_0 - \beta_1$ so $p - q = k - m(G)$ and we see that Corollary 4.5(b) is the Euler-Poincaré equation for graphs.

MATROIDS

This subject was first introduced by Whitney [W15]. A discussion of the basic properties of matroids, as well as several equivalent axiomatic formulations, may be found in Whitney's original paper.

A *matroid* consists of a finite set M of elements together with a family $\mathscr{C} = \{C_1, C_2, \cdots\}$ of nonempty subsets of M, called *circuits*, satisfying the axioms:

1. no proper subset of a circuit is a circuit;
2. if $x \in C_1 \cap C_2$, then $C_1 \cup C_2 - \{x\}$ contains a circuit.

With every graph G, one can associate a matroid by taking its set X of lines as the set M, and its cycles as the circuits. It is easily seen that the two axioms are satisfied. It is slightly less obvious that G yields another matroid by taking the cocycles of G as the circuits. These are called respectively the *cycle matroid* and the *cocycle matroid* of G.

Another, equivalent, definition of matroid is as follows. A *matroid* consists of a finite set M of elements together with a family of subsets of M called *independent sets* such that:

1. the empty set is independent;

2. every subset of an independent set is independent;

3. for every subset A of M, all maximal independent sets contained in A have the same number of elements.

A graph G yields a matroid in this sense by taking the lines of G as set M and the acyclic subgraphs of G as the independent sets.

The duality (cycles vs. cocycles, trees vs. cotrees) which appears in the preceding section is closely related to duality in matroids. Minty [M12] constructed a self-dual axiom system for "graphoids" which displays matroid duality explicitly.

A *graphoid* consists of a set M of elements together with two collections $\mathscr{C}$ and $\mathscr{D}$ of nonempty subsets of M, called *circuits* and *cocircuits* respectively, such that:

1. for any $C \in \mathscr{C}$ and $D \in \mathscr{D}$, $|C \cap D| \neq 1$;

2. no circuit properly contains another circuit and no cocircuit properly contains another cocircuit;

3. for any painting of M which colors exactly one element green and the rest either red or blue, there exists either

   a) a circuit C containing the green element and no red elements, or

   b) a cocircuit D containing the green element and no blue elements.

While the cycles of every graph form a matroid, not every matroid can so arise from a graph, as we shall see in Chapter 13. Two comprehensive references on matroid theory are Minty [M12] and Tutte [T19].

Excursion

Ulam's conjecture is still as unsolved as ever for arbitrary graphs. But Kelly [K5] proved its validity for trees. As we have seen, the point of view toward this conjecture proposed in [H29] is that if G has $p \geq 3$ and one is presented with the p unlabeled subgraphs $G_i = G - v_i$, then the graph G itself can be reconstructed uniquely from the G_i. Kelly's result for trees was extended in [HP6] where it is shown that every nontrivial tree T can be reconstructed from only those subgraphs $T_i = T - v_i$ which are themselves trees, that is, such that v_i is an endpoint. This has been improved, in turn, by Bondy, who showed [B15] that a tree T can be reconstructed from its subgraphs $T - v_i$ with the v_i the *peripheral* points, those whose eccentricity equals the diameter of T. Manvel [M2] then showed that almost* every tree T can be reconstructed using only those subtrees $T - v_i$ which are non-isomorphic. Another class of graphs has been reconstructed by Manvel [M3], namely *unicyclic* graphs, which are connected and have just one cycle.

---

* With just two pairs of exceptional trees.

EXERCISES

4.1 Draw all trees with nine points. Then compare your diagrams with those in Appendix II.

4.2 Every tree is a bigraph. Which trees are complete bigraphs?

4.3 The following four statements are equivalent.

(1) G is a forest.
(2) Every line of G is a bridge.
(3) Every block of G is K_2.
(4) Every nonempty intersection of two connected subgraphs of G is connected.

4.4 The following four statements are equivalent.

(1) G is unicyclic.
(2) G is connected and $p = q$.
(3) For some line x of G, the graph $G - x$ is a tree.
(4) G is connected and the set of lines of G which are not bridges form a cycle.

(Anderson and Harary [AH1])

4.5 For any connected graph G, $r(G) \le d(G) \le 2r(G)$.

4.6 Construct a tree with disjoint center and centroid, each having two points.

4.7 The center of any connected graph lies in a block. (Harary and Norman [HN2])

4.8 Given the block-cutpoint tree $bc(G)$ of a connected graph G, determine the block-graph $B(G)$ and the cutpoint-graph $C(G)$.

4.9 Determine the cycle ranks of (a) K_p, (b) $K_{m,n}$, (c) a connected cubic graph with p points.

4.10 The intersection of a cycle and a cocycle contains an even number of lines.

4.11 A graph is bipartite if and only if every cycle in some cycle basis is even.

4.12 Every connected graph has a spanning tree.

4.13 Show how the block-cutpoint graph of any graph can be defined as an intersection graph.

4.14 A cotree of a connected graph is a maximal subgraph containing no cocycles.

4.15 A tree with $p \ge 3$ has diameter 2 if and only if it is a star.

4.16 Prove or disprove:

   a) If G has diameter 2, then it has a spanning star.
   b) If G has a spanning star, then it has diameter 2.

4.17 Determine all connected graphs G for which $G \cong bc(G)$.

*4.18 The maximum number of lines in a graph with p points and radius r is

$$\binom{p}{2} \quad \text{if} \quad r = 1,$$

$$[p(p - 2)/2] \quad \text{if} \quad r = 2,$$

$$\tfrac{1}{2}(p^2 - 4rp + 5p + 4r^2 - 6r) \quad \text{if} \quad r \ge 3. \qquad \text{(Vizing [V5])}$$

4.19 G is a block if and only if every two lines lie on a common cocycle.

CHAPTER 5

CONNECTIVITY

> We must all hang together,
> or assuredly we shall all hang separately.
>
> B. Franklin

The connectivity of graphs is a particularly intuitive area of graph theory and extends the concepts of cutpoint, bridge, and block. Two invariants called connectivity and line-connectivity are useful in deciding which of two graphs is "more connected."

There is a rich body of theorems concerning connectivity. Many of these are variations of a classical result of Menger, which involves the number of disjoint paths joining a given pair of points in a graph. We will see that several such variations have been discovered in areas of mathematics other than graph theory.

CONNECTIVITY AND LINE-CONNECTIVITY

The *connectivity* $\kappa = \kappa(G)$ of a graph G is the minimum number of points whose removal results in a disconnected or trivial graph. Thus the connectivity of a disconnected graph is 0, while the connectivity of a connected graph with a cutpoint is 1. The complete graph K_p cannot be disconnected by removing any number of points, but the trivial graph results after removing $p - 1$ points; therefore, $\kappa(K_p) = p - 1$. Sometimes κ is called the *point-connectivity*.

Analogously, the *line-connectivity* $\lambda = \lambda(G)$ of a graph G is the minimum number of lines whose removal results in a disconnected or trivial graph. Thus $\lambda(K_1) = 0$ and the line-connectivity of a disconnected graph is 0, while that of a connected graph with a bridge is 1. Connectivity, line-connectivity, and minimum degree are related by an inequality due to Whitney [W11].

Theorem 5.1 For any graph G,

$$\kappa(G) \leq \lambda(G) \leq \delta(G).$$

Proof. We first verify the second inequality. If G has no lines, then $\lambda = 0$. Otherwise, a disconnected graph results when all the lines incident with a point of minimum degree are removed. In either case, $\lambda \leq \delta$.

To obtain the first inequality, various cases are considered. If G is disconnected or trivial, then $\kappa = \lambda = 0$. If G is connected and has a bridge x, then $\lambda = 1$. In this case, $\kappa = 1$ since either G has a cutpoint incident with x or G is K_2. Finally, suppose G has $\lambda \geq 2$ lines whose removal disconnects it. Clearly, the removal of $\lambda - 1$ of these lines produces a graph with a bridge $x = uv$. For each of these $\lambda - 1$ lines, select an incident point different from u or v. The removal of these points also removes the $\lambda - 1$ lines and quite possibly more. If the resulting graph is disconnected, then $\kappa < \lambda$; if not, x is a bridge, and hence the removal of u or v will result in either a disconnected or a trivial graph, so $\kappa \leq \lambda$ in every case. (See Fig. 5.1.)

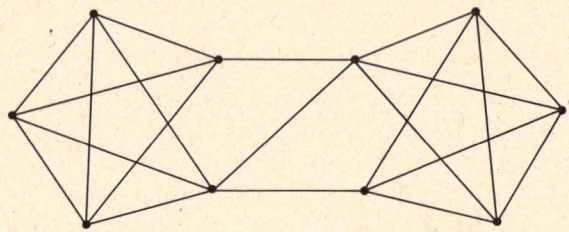

Fig. 5.1. A graph for which $\kappa = 2$, $\lambda = 3$, and $\delta = 4$.

Chartrand and Harary [CH4] constructed a family of graphs with prescribed connectivities which also have a given minimum degree. This result shows that the restrictions on κ, λ, and δ imposed by Theorem 5.1 cannot be improved.

Theorem 5.2 For all integers a, b, c such that $0 < a \leq b \leq c$, there exists a graph G with $\kappa(G) = a$, $\lambda(G) = b$, and $\delta(G) = c$.

Chartrand [C8] pointed out that if δ is large enough, then the second inequality of Theorem 5.1 becomes an equality.

Theorem 5.3 If G has p points and $\delta(G) \geq [p/2]$, then $\lambda(G) = \delta(G)$.

For example, if G is regular of degree $r \geq p/2$, then $\lambda(G) = r$. In particular, $\lambda(K_p) = p - 1$.

The analogue of Theorem 5.3 for connectivity does not hold. The problem of determining the largest connectivity possible for a graph with a given number of points and lines was proposed by Berge [B11] and a solution was given in [H26].

Theorem 5.4 Among all graphs with p points and q lines, the maximum connectivity is 0 when $q < p - 1$ and is $[2q/p]$, when $q \geq p - 1$.

Outline of proof.   Since the sum of the degrees of any (p, q) graph G is $2q$, the mean degree is $2q/p$. Therefore $\delta(G) \leq [2q/p]$, so $\kappa(G) \leq [2q/p]$ by Theorem 5.1.   To show that this value can actually be attained, an appropriate family of graphs can be constructed.   The same construction also gives those (p, q) graphs with maximum line-connectivity.

Corollary 5.4(a)   The maximum line-connectivity of a (p, q) graph equals the maximum connectivity.

Only very recently the question of separating a graph by removing a mixed set of points and lines has been studied.   A *connectivity pair* of a graph G is an ordered pair (a, b) of nonnegative integers such that there is some set of a points and b lines whose removal disconnects the graph and there is no set of $a - 1$ points and b lines or of a points and $b - 1$ lines with this property.   Thus in particular the two ordered pairs $(\kappa, 0)$ and $(0, \lambda)$ are connectivity pairs for G, so that the concept of connectivity pair generalizes both the point-connectivity and the line-connectivity of a graph.   It is readily seen that for each value of a, $0 \leq a \leq \kappa$, there is a unique connectivity pair (a, b_a); thus G has exactly $\kappa + 1$ connectivity pairs.

The connectivity pairs of a graph G determine a function f from the set $\{0, 1, \cdots, \kappa\}$ into the nonnegative integers such that $f(\kappa) = 0$ (cf. Theorem 5.1).   This is called the *connectivity function* of G.   It is strictly decreasing, since if (a, b) is a connectivity pair with $b > 0$ there is obviously a set of $a + 1$ points and $b - 1$ lines whose removal disconnects the graph or leaves only one point.   The following theorem, proved by construction in Beineke and Harary [BH6], shows that these are the only conditions which a connectivity function must satisfy.

Theorem 5.5   Every decreasing function f from $\{0, 1, \cdots, \kappa\}$ into the nonnegative integers such that $f(\kappa) = 0$ is the connectivity function of some graph.

A graph G is *n-connected* if $\kappa(G) \geq n$ and *n-line-connected* if $\lambda(G) \geq n$. We note that a nontrivial graph is 1-connected if and only if it is connected, and that it is 2-connected if and only if it is a block having more than one line.   So K_2 is the only block not 2-connected.   From Theorem 3.3, it therefore follows that G is 2-connected if and only if every two points of G lie on a cycle.   Dirac [D8] extended this observation to *n*-connected graphs.

Theorem 5.6   If G is *n*-connected, $n \geq 2$, then every set of n points of G lie on a cycle.

By taking G to be the cycle C_n itself, it is seen that the converse is not true for $n > 2$.

A characterization of 3-connected graphs also exists, although its formulation is not as easily given.   In order to present this result, we need

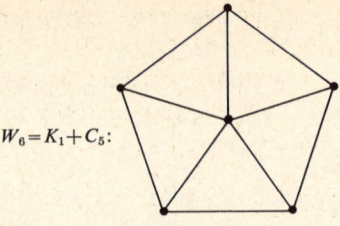

$W_6 = K_1 + C_5$:

Fig. 5.2. A wheel.

the "wheel" invented by the eminent graph theorist W. T. Tutte. For $n \geq 4$, the wheel W_n is defined to be the graph $K_1 + C_{n-1}$. (See Fig. 5.2.)

Tutte's theorem [T13] characterizing 3-connected graphs can now be stated.

Theorem 5.7 A graph G is 3-connected if and only if G is a wheel or can be obtained from a wheel by a sequence of operations of the following two types:

1. The addition of a new line.
2. The replacement of a point v having degree at least 4 by two adjacent points v', v'' such that each point formerly joined to v is joined to exactly one of v' and v'' so that in the resulting graph, deg $v' \geq 3$ and deg $v'' \geq 3$.

The graph G of Fig. 5.3 is 3-connected since it can be obtained from the wheel W_5 as indicated.

An *n-component* of a graph G is a maximal *n*-connected subgraph. In particular, the 1-components of G are the nontrivial components of G while the 2-components are the blocks of G with at least 3 points. It is readily seen that two different 1-components have no points in common, and two

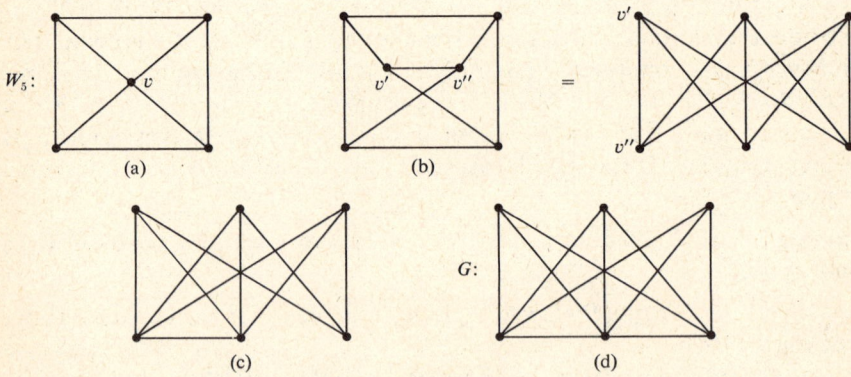

Fig. 5.3. Demonstration that a graph is 3-connected.

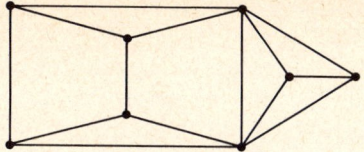

Fig. 5.4. A graph with two 3-components which meet in two points.

distinct 2-components meet in at most one point. These facts have been generalized by Harary and Kodama [HK1]. (See Fig. 5.4.)

Theorem 5.8 Two distinct n-components of a graph G have at most $n - 1$ points in common.

GRAPHICAL VARIATIONS OF MENGER'S THEOREM

In 1927 Menger [M9] showed that the connectivity of a graph is related to the number of disjoint paths joining distinct points in the graph. Many of the variations and extensions of Menger's result which have since appeared have been graphical, and we discuss some of these here. By emphasizing the form these theorems take, it is possible to classify them in an illuminating way.

Let u and v be two distinct points of a connected graph G. Two paths joining u and v are called *disjoint* (sometimes called point-disjoint) if they have no points other than u and v (and hence no lines) in common; they are *line-disjoint* if they have no lines in common. A set S of points, lines, or points and lines *separates* u and v if u and v are in different components of $G - S$. Clearly, no set of points separates two adjacent points. Menger's Theorem was originally presented in the "point form" given in Theorem 5.9.

Theorem 5.9 The minimum number of points separating two nonadjacent points s and t is the maximum number of disjoint s–t paths.

Proof. We follow the elegant proof of Dirac [D10]. It is clear that if k points separate s and t, then there can be no more than k disjoint paths joining s and t.

It remains to show that if it takes k points to separate s and t in G, there are k disjoint s–t paths in G. This is certainly true if $k = 1$. Assume it is not true for some $k > 1$. Let h be the smallest such k, and let F be a graph with the minimum number of points for which the theorem fails for h. We remove lines from F until we obtain a graph G such that h points are required to separate s and t in G but for any line x of G, only $h - 1$ points are required to separate s and t in $G - x$. We first investigate the properties of this graph G, and then complete the proof of the theorem.

By the definition of G, for any line x of G there exists a set $S(x)$ of $h - 1$ points which separates s and t in $G - x$. Now $G - S(x)$ contains at least one s–t path, since it takes h points to separate s and t in G. Each such s–t path must contain the line $x = uv$ since it is not a path in $G - x$. So $u, v \notin S(x)$ and if $u \neq s, t$ then $S(x) \cup \{u\}$ separates s and t in G.

If there is a point w adjacent to both s and t in G, then $G - w$ requires $h - 1$ points to separate s and t and so it has $h - 1$ disjoint s–t paths. Replacing w, we have h disjoint s–t paths in G. So we have shown:

(I) No point is adjacent to both s and t in G.

Let W be any collection of h points separating s and t in G. An s–W path is a path joining s with some $w_i \in W$ and containing no other point of W. Call the collections of all s–W paths and W–t paths P_s and P_t respectively. Then each s–t path begins with a member of P_s and ends with a member of P_t, because every such path contains a point of W. Moreover, the paths in P_s and P_t have the points of W and no others in common, since it is clear that each w_i is in at least one path in each collection and, if some other point were in both an s–W and a W–t path, then there would be an s–t path containing no point of W. Finally, either $P_s - W = \{s\}$ or $P_t - W = \{t\}$, since, if not, then both P_s plus the lines $\{w_1 t, w_2 t, \cdots\}$ and P_t plus the lines $\{sw_1, sw_2, \cdots\}$ are graphs with fewer points than G in which s and t are nonadjacent and h-connected, and therefore in each there are h disjoint s–t paths. Combining the s–W and W–t portions of these paths, we can construct h disjoint s–t paths in G, and thus have a contradiction. Therefore we have proved:

(II) Any collection W of h points separating s and t is adjacent either to s or to t.

Now we can complete the proof of the theorem. Let $P = \{s, u_1, u_2, \cdots, t\}$ be a shortest s–t path in G and let $u_1 u_2 = x$. Note that by (I), $u_2 \neq t$. Form $S(x) = \{v_1, v_2, \cdots, v_{h-1}\}$ as above, separating s and t in $G - x$. By (I), $u_1 t \notin G$, so by (II), with $W = S(x) \cup \{u_1\}$, $sv_i \in G$, for all i. Thus by (I), $v_i t \notin G$, for all i. However, if we pick $W = S(x) \cup \{u_2\}$ instead, we have by (II) that $su_2 \in G$, contradicting our choice of P as a shortest s–t path, and completing the proof of the theorem.

In Fig. 5.5 we display a graph with two nonadjacent points s and t which can be separated by removing three points but no fewer. In accordance with the theorem, the maximum number of disjoint s–t paths is 3.

Chronologically the second variation of Menger's Theorem was published by Whitney in a paper [W11] in which he included a criterion for a graph to be n-connected.

Theorem 5.10 A graph is n-connected if and only if every pair of points are joined by at least n point-disjoint paths.

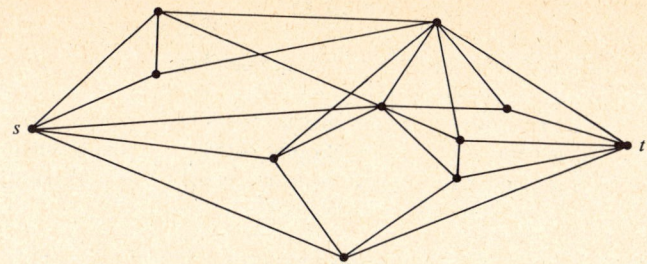

Fig. 5.5. A graph illustrating Menger s Theorem.

An indication of the relationship between Theorems 5.9 and 5.10 is easily supplied by introducing the concept of local connectivity. The *local connectivity* of two nonadjacent points u and v of a graph is denoted by $\kappa(u, v)$ and is defined as the smallest number of points whose removal separates u and v. In these terms, Menger's Theorem asserts that for any two specific nonadjacent points u and v, $\kappa(u, v) = \mu_0(u, v)$, the maximum number of point-disjoint paths joining u and v. Obviously both theorems hold for complete graphs. If we are dealing with a graph G which is not complete, then the observation which links Theorems 5.9 and 5.10 is that $\kappa(G) = \min \kappa(u, v)$ over all pairs of nonadjacent points u and v.

Strangely enough, the theorem analogous to Theorem 5.9 in which the pair of points are separated by a set of lines was not discovered until much later. There are several nearly simultaneous discoveries of this result which appeared in papers by Ford and Fulkerson [FF1] (as a special case of their "max-flow, min-cut" theorem) and Elias, Feinstein, and Shannon [EFS1], and also in unpublished work of A. Kotzig.

Theorem 5.11   For any two points of a graph, the maximum number of line-disjoint paths joining them equals the minimum number of lines which separate them.

Referring again to Fig. 5.5, we see that s and t can be separated by the removal of five lines but no fewer, and that the maximum number of line-disjoint s–t paths is five.

Even with only these three theorems available, we can see the beginnings of a scheme for classifying them. The difference between Theorems 5.9 and 5.10 may be expressed by saying that Theorem 5.9 involves two specific points of a graph while Theorem 5.10 gives a bound in terms of two general points. This distinction, as well as the obvious one between Theorems 5.9 and 5.11, is indicated in Table 5.1.

Thus we see that with no additional effort we can get another variation of Menger's Theorem by stating the line form of the Whitney result.

Table 5.1

Theorem	Objects separated	Maximum number	Minimum number
5.9	specific u, v	disjoint paths	points separating u, v
5.10	general u, v	disjoint paths	points separating u, v
5.11	specific u, v	line-disjoint paths	lines separating u, v

Theorem 5.12 A graph is n-line-connected if and only if every pair of points are joined by at least n line-disjoint paths.

In Menger's original paper there also appeared the following variation involving sets of points rather than individual points.

Theorem 5.13 For any two disjoint nonempty sets of points V_1 and V_2, the maximum number of disjoint paths joining V_1 and V_2 is equal to the minimum number of points which separate V_1 and V_2.

Of course it must be specified that no point of V_1 is adjacent with a point of V_2 for the same reason as in Theorem 5.9. Two paths joining V_1 and V_2 are understood to be disjoint if they have no points in common other than their endpoints. A proof of the equivalence of Theorems 5.9 and 5.13 is extremely straightforward and only involves shrinking the sets of points V_1 and V_2 to individual points.

Another variation is given in the next theorem, considered by Dirac [D9]. Because the proof involves typical methods in the demonstration of equivalence of these variations, we include it in full.

Theorem 5.14 A graph with at least $2n$ points is n-connected if and only if for any two disjoint sets V_1 and V_2 of n points each, there exist n disjoint paths joining these two sets of points.

Note that in this theorem these n disjoint paths do not have any points at all in common, not even their endpoints!

Proof. To show the sufficiency of the condition, we form the graph G' from G by adding two new points w_1 and w_2 with w_i adjacent to exactly the points of V_i, $i = 1, 2$. (See Fig. 5.6.)

Since G is n-connected, so is G', and hence by Theorem 5.9 there are n disjoint paths joining w_1 and w_2. The restrictions of these paths to G are clearly the n disjoint V_1–V_2 paths we need.

To prove the other "half," let S be a set of at least $n - 1$ points which separates G into G_1 and G_2, with points sets V'_1 and V'_2 respectively. Then, since $|V'_1| \geq 1$, $|V'_2| \geq 1$, and $|V'_1| + |V'_2| + |S| = |V| \geq 2n$, there is a partition of S into two disjoint subsets S_1 and S_2 such that $|V'_1 \cup S_1| \geq n$ and $|V'_2 \cup S_2| \geq n$. Picking any n-subsets V_1 of $V'_1 \cup S_1$, and V_2 of $V'_2 \cup S_2$,

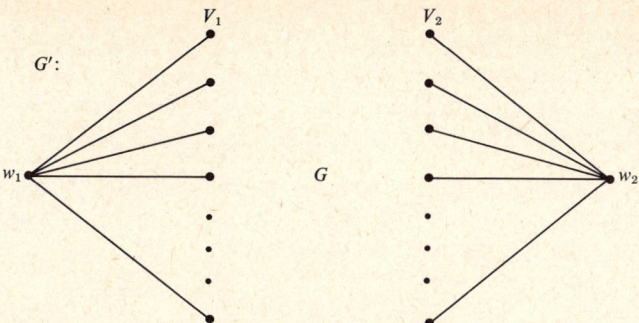

Fig. 5.6. Construction of G'.

we have two disjoint sets of n points each. Every path joining V_1 and V_2 must contain a point of S, and since we know there are n disjoint V_1–V_2 paths, we see that $|S| \geq n$, and G is n-connected.

We have defined connectivity pairs for a graph. Similarly, one can define connectivity pairs for two specific points u and v. It is then natural to ask for a mixed form of Menger's Theorem involving connectivity pairs. The following theorem of Beineke and Harary [BH6] is one such result; a proof can be readily supplied by imitating that of Theorem 5.9.

Theorem 5.15 The ordered pair (a, b) is a connectivity pair for points u and v in a graph G if and only if there exist a point-disjoint u–v paths and also b line-disjoint u–v paths which are line-disjoint from the preceding a paths, and further these are the maximum possible numbers of such paths.

In general, all of the theorems we have mentioned have corresponding digraph forms, and in fact Dirac points out that his proof of Menger's Theorem works equally well for directed graphs. At this point, then, we could add eleven more theorems to Table 5.1, namely Theorems 5.12 through 5.15, and the directed forms of Theorems 5.9 through 5.15. This would be a somewhat futile effort, however, since it should be clear that the table would still be far from complete. To count the total number of variations which have been suggested up to this point, we note that we may consider either a graph G or a digraph D, in which we may separate

  i) specific points u, v,
 ii) general points u, v,
iii) two sets of points V_1, V_2 (as in Theorem 5.13).

This separation may be accomplished by removing

  i) points,
 ii) lines, or
iii) points and lines (as in Theorem 5.15).

By taking all possible combinations of these alternatives, we could construct $2 \cdot 3 \cdot 3 = 18$ theorems. The fact that all of these theorems are true may be verified by the reader, although it would be a tedious exercise.

Finally, Fulkerson [F13] proved the following theorem, which deals with disjoint cutsets instead of disjoint paths.

Theorem 5.16 In any graph, the maximum number of line-disjoint cutsets of lines separating two points u and v is equal to the minimum number of lines in a path joining u and v; that is, to $d(u, v)$.

Although this theorem is of Mengerian type, it is much easier to prove than Menger's Theorem. By taking all the possible variations of this theorem, as we have with the theorems involving paths, we could increase the number of Mengerian theorems again.

FURTHER VARIATIONS OF MENGER'S THEOREM

In this section we include several additional variations of Menger's Theorem, all discovered independently and only later seen to be related to each other and to a graph theoretic formulation.

A *network N* may be regarded as a graph or directed graph together with a function which assigns a positive real number to each line. For precise definitions of "maximum flow" and "minimum cut capacity," see the book [FF2] by Ford and Fulkerson.

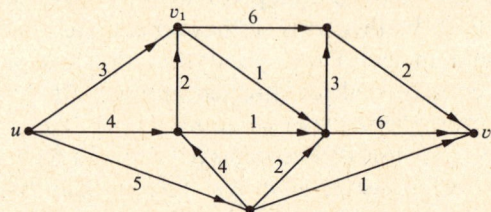

Fig. 5.7. A network with integral capacities.

Theorem 5.17 In any network N in which there is a path from u to v, the maximum flow from u to v is equal to the minimum cut capacity.

It is straightforward but not entirely obvious to verify that in Fig. 5.7 the maximum flow in the network from u to v is 7, and that the minimum cut capacity is also 7.

In the case where all the capacities are positive integers, as in this network, there is an immediate equivalence between the maximum flow theorem and that variation of Menger's Theorem in which the setting is a directed multigraph D and there are two specific points u and v. The transformation

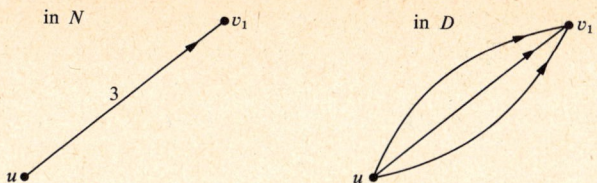

Fig. 5.8.   The transformation from network to multigraph.

which makes this equivalence apparent is displayed in Fig. 5.8 in which the directed line from u to v_1 in Fig. 5.7 which has capacity 3 is transformed into three directed lines without any capacity indicated.

Let us define a *line of a matrix* as either a row or a column. In a binary matrix M, a collection of lines is said to *cover* all the unit entries of M if every 1 is in one of these lines. Two 1's of M are called *independent* if they are neither in the same row nor in the same column. König [K9] obtained the next variation of Menger's Theorem in these terms; compare Theorem 10.2.

Theorem 5.18   In any binary matrix, the maximum number of independent unit elements equals the minimum number of lines which cover all the units.

$$M = \begin{bmatrix} 0 & 0 & 1 & 0 & 0 & 0 \\ 1 & 1 & 0 & 1 & 0 & 1 \\ 0 & 0 & 1 & 0 & 0 & 1 \\ 0 & 1 & 1 & 0 & 1 & 0 \\ 0 & 0 & 1 & 0 & 0 & 1 \end{bmatrix} \qquad M' = \begin{bmatrix} 0 & 0 & 1 & 0 & 0 & 0 \\ 1 & 0 & 0 & 0 & 0 & 0 \\ 0 & 0 & 0 & 0 & 0 & 1 \\ 0 & 1 & 0 & 0 & 0 & 0 \\ 0 & 0 & 0 & 0 & 0 & 0 \end{bmatrix}$$

We illustrate Theorem 5.18 with the binary matrix M above. All the unit entries of M are covered by rows 2 and 4 and columns 3 and 6, but there is no collection of three lines of M which covers all its 1's. In the matrix M' there are shown four independent unit entries of M and there is no set of five independent 1's in M.

When this matrix M is regarded as an incidence matrix of sets versus elements, Theorem 5.18 becomes very closely related to the celebrated theorem of P. Hall [H8], which provides a criterion for a collection of finite sets $S_1, S_2, \cdots, S_m$ to possess a system of distinct representatives. This means a set $\{e_1, e_2, \cdots, e_m\}$ of distinct elements such that e_i is in S_i, for each i. We present here the proof of Hall's Theorem which is due to Rado [R1].

Theorem 5.19   There exists a system of distinct representatives for a family of sets $S_1, S_2, \cdots, S_m$ if and only if the union of any k of these sets contains at least k elements, for all k from 1 to m.

Proof.   The necessity is immediate. For the sufficiency we first prove that if the collection $\{S_i\}$ satisfies the stated conditions and $|S_m| \geq 2$, then there is an element e in S_m such that the collection of sets $S_1, S_2, \cdots, S_{m-1}, S_m - \{e\}$

also satisfies the conditions. Suppose this is not the case. Then there are elements e and f in S_m and subsets J and K of $\{1, 2, \cdots, m - 1\}$ such that

$$\left|\left(\bigcup_{i \in J} S_i\right) \cup (S_m - \{e\})\right| < |J| + 1 \quad \text{and} \quad \left|\left(\bigcup_{i \in K} S_i\right) \cup (S_m - \{f\})\right| < |K| + 1.$$

But then

$$|J| + |K| \geq \left|\left(\bigcup_J S_i\right) \cup (S_m - \{e\})\right| + \left|\left(\bigcup_K S_i\right) \cup (S_m - \{f\})\right|$$

$$\geq \left|\left(\bigcup_{J \cup K} S_i\right) \cup S_m\right| + \left|\bigcup_{J \cap K} S_i\right|$$

$$\geq |J \cup K| + 1 + |J \cap K| > |J| + |K|,$$

which is a contradiction.

The sufficiency now follows by induction on the maximum of the numbers $|S_i|$. If each set is a singleton, there is nothing to prove. The induction step is made by application (repeated if necessary) of the above result to the sets of largest order.

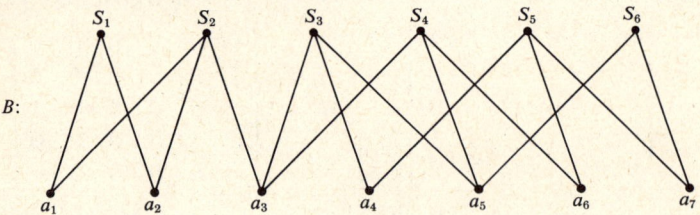

Fig. 5.9. A bipartite graph illustrating Hall's Theorem.

In Fig. 5.9 we show a bipartite graph B in which the points refer either to sets S_i or to elements a_j. Two points of B are adjacent if and only if one is a set point, the other is an element point, and the element is a member of the set. The link between Theorem 5.19 and Menger's Theorem is accomplished by introducing two new points into a graph of the form of Fig. 5.9. Call these points u and v and join u to every set point S_i and v with every element point a_j to obtain a new graph. Theorem 5.19 can then be proved by applying either the maximum flow theorem or the appropriate line form of Menger's Theorem to this graph.

Although the following theorem due to Dilworth [D4] is expressed in terms of lattice theory,* it has been established (see Mirsky and Perfect [MP1]) that the result is equivalent to Hall's Theorem. Two elements of a lattice (see Birkhoff [B13]) are incomparable if neither dominates the other.

---

* More generally the result holds for partially ordered sets.

By a chain in a lattice is meant a downward path from an upper element to a lower element in the "Hasse diagram" of the lattice.

Theorem 5.20 In any finite lattice, the maximum number of incomparable elements equals the minimum number of chains which include all the elements.

For example, in the lattice of the 3-cube, there are at most three incomparable elements; it is easy to cover all the elements with three chains but impossible to do so with only two chains.

We have seen in this section several theorems of Mengerian type occurring in settings which are not graph theoretic. A more extensive treatment of such results appears in the review article [H33]. For an elegant summary of the vast literature on theorems involving systems of distinct representatives, see Mirsky and Perfect [MP1].

EXERCISES

5.1 The connectivity of

    a) the octahedron $\bar{K}_2 + C_4$ is 4.
    b) the square of a polygon C_n, $n \geq 5$, is 4.

5.2 Every n-connected graph has at least $pn/2$ lines.

5.3 Construct a graph with $\kappa = 3$, $\lambda = 4$, $\delta = 5$.

5.4 Theorem 5.3 does not hold if $\lambda(G)$ is replaced by $\kappa(G)$.

5.5 There exists no 3-connected graph with seven lines.

5.6 The connectivity and line-connectivity are equal in every cubic graph.

5.7 Determine which connectivity pairs can occur in 4-regular graphs.

5.8 If G is regular of degree r and $\kappa = 1$, then $\lambda \leq [r/2]$.

5.9 Construct a family of (p, q) graphs with $2q/p$ integral such that $\kappa = 2q/p$.

5.10 Let G be a complete n-partite graph other than C_4. Then every minimum line cutset is the coboundary of some point.   (M. D. Plummer)

5.11 Find the connectivity function for s and t in the graph of Fig. 5.5.

5.12 Find a graph with points s and t for which the connectivity function is (0, 5), (1, 3), (2, 2), (3, 0).

5.13 Use Tutte's Theorem 5.7 to show that the graph of the cube Q_3 is 3-connected.

5.14 Every block of a connected graph G is a wheel if and only if $q = 2p - 2$ and $\kappa(u, v) = 1$ or 3 for any two nonadjacent points u, v.   (Bollobás [B14])

5.15 Every cubic triply-connected graph can be obtained from K_4 by the following construction. Replace two distinct lines $u_1 v_1$ and $u_2 v_2$ ($u_1 = u_2$ is permitted) by the subgraph with two new points w_1, w_2 and the new lines $u_1 w_1$, $w_1 v_1$, $u_2 w_2$, $w_2 v_2$, and $w_1 w_2$.

5.16 Given two disjoint paths P_1 and P_2 joining two points u and v of a 3-connected graph G, is it always possible to find a third path joining u and v which is disjoint from both P_1 and P_2?

5.17 State the result analogous to Theorem 5.9 for the maximum number of disjoint paths joining two adjacent points of a graph.

*5.18 If $f_r(p)$ is the smallest number such that for $q \geq f_r(p)$ every (p, q) graph has two points joined by r disjoint paths, then

$$f_2(p) = p, \qquad f_3(p) = [(3p - 1)/2], \qquad \text{and} \qquad f_4(p) = 2p - 1.$$

(Bollobás [B14])

5.19 If G has diameter d and $\kappa \geq 1$, then $p \geq \kappa(d - 1) + 2$.   (Watkins [W5])

5.21 If G is connected, then

$$\kappa(G) = 1 + \min_{v \in V} \kappa(G - v)$$

5.22 In any graph, the maximum number of disjoint cutsets of points separating two points u and v equals $d(u, v) - 1$.

5.23 In a κ-minimal graph G, $\kappa(G - x) < \kappa(G)$ for every line x.

  a) G is κ-minimal if and only if $\kappa(u, v) = \kappa(G)$ for every pair of adjacent points u, v.
  b) If G is κ-minimal then $\delta = \kappa$.   (Halin [H5])

5.24 Prove the equivalence of Theorems 5.18 and 5.19. (See for example M. Hall [H7, p. 49]).

5.25 If G is n-connected, $n \geq 2$, and $\delta(G) \geq (3n - 1)/2$, then there exists a point v in G such that $G - v$ is n-connected.   (Chartrand, Kaugars, and Lick [CKL1])

CHAPTER 6

PARTITIONS

Gallia est omnis divisa in tres partes.

JULIUS CAESAR, *de Bello Gallico*

The degrees $d_1, \cdots, d_p$ of the points of a graph form a sequence of non-negative integers, whose sum is of course $2q$. In number theory it is customary to define a partition of a positive integer n as a list or unordered sequence of positive integers whose sum is n. Under this definition, 4 has five partitions:

$$4, \quad 3 + 1, \quad 2 + 2, \quad 2 + 1 + 1, \quad 1 + 1 + 1 + 1.$$

The order of the summands in a partition is not important. The degrees of a graph with no isolated points determine such a partition of $2q$, but because of the importance of having a general definition holding for all graphs, it is convenient to use an extended definition, changing positive to nonnegative.

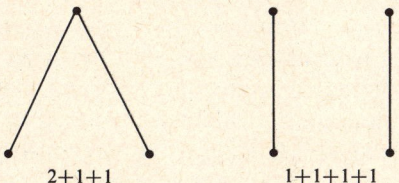

2+1+1   1+1+1+1

Fig. 6.1. The graphical partitions of 4.

A *partition of a nonnegative integer n* is a finite list of nonnegative integers with sum n. In this sense, the partitions of 4 also allow an arbitrary finite number of zero summands. The *partition of a graph* is the partition of $2q$ as the sum of the degrees of the points, $2q = \Sigma\, d_i$, as in Theorem 2.1. Only two of the five partitions of 4 into positive summands belong to a graph, see Fig. 6.1.

A partition $\Sigma\, d_i$ of n into p parts is *graphical* if there is a graph G whose points have degrees d_i. If such a partition is graphical, then certainly every

$d_i \leq p - 1$, and n is even. These two conditions are not sufficient for a partition to be graphical, as shown by the partition $10 = 3 + 3 + 3 + 1$. Two related questions arise. First, how can one tell whether a given partition is graphical? Second, how can one construct a graph for a given graphical partition? An existential answer to the first was given by Erdös and Gallai [EG1]. Another answer found independently by Havel [H36] and Hakimi [H4] is constructive in nature, and so answers the second question as well. We first give this result.

Theorem 6.1 A partition $\Pi = (d_1, d_2, \cdots, d_p)$ of an even number into p parts with $p - 1 \geq d_1 \geq d_2 \geq \cdots \geq d_p$ is graphical if and only if the modified partition

$$\Pi' = (d_2 - 1, d_3 - 1, \cdots, d_{d_1+1} - 1, d_{d_1+2}, \cdots, d_p)$$

is graphical.

Proof. If Π' is graphical, then so is Π, since from a graph with partition Π' one can construct a graph with partition Π by adding a new point adjacent to points of degrees $d_2 - 1, d_3 - 1, \cdots, d_{d_1+1} - 1$.

Now let G be a graph with partition Π. If a point of degree d_1 is adjacent to points of degrees d_i for $i = 2$ to $d_1 + 1$, then the removal of this point results in a graph with partition Π'.

Therefore we will show that from G one can get a graph with such a point. Suppose that G has no such point. We assume that in G, point v_i has degree d_i, with v_1 being a point of degree d_1 for which the sum of the degrees of the adjacent points is maximum. Then there are points v_i and v_j with $d_i > d_j$ such that $v_1 v_j$ is a line but $v_1 v_i$ is not. Therefore some point v_k is adjacent to v_i but not to v_j. Removal of the lines $v_1 v_j$ and $v_k v_i$ and addition of $v_1 v_i$ and $v_k v_j$ results in another graph with partition Π in which the sum of the degrees of the points adjacent to v_1 is greater than before. Repeating this process results in a graph in which v_1 has the desired property.

The theorem gives an effective algorithm for constructing a graph with a given partition, if one exists. If none exists, the algorithm cannot be applied at some step.

Corollary 6.1 (Algorithm) A given partition $\Pi = (d_1, d_2, \cdots, d_p)$ with

$$p - 1 \geq d_1 \geq d_2 \geq \cdots \geq d_p$$

is graphical if and only if the following procedure results in a partition with every summand zero.

1. Determine the modified partition Π' as in the statement of Theorem 6.1.
2. Reorder the terms of Π' so that they are nonincreasing, and call the resulting partition Π_1.

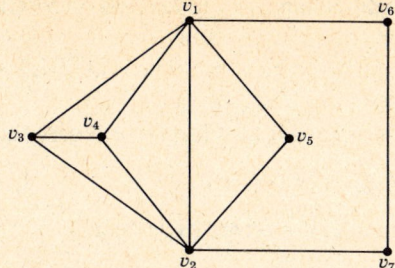

Fig. 6.2. An example of the algorithm for graphical partitions.

3. Determine the modified partition Π'' of Π_1 as in step 1, and the re-ordered partition Π_2.

4. Continue the process as long as nonnegative summands can be obtained.

If a partition obtained at an intermediate stage is known to be graphical, stop, since Π itself is then established as graphical. To illustrate this algorithm, we test the partition

$$\Pi = (5, 5, 3, 3, 2, 2, 2)$$
$$\Pi' = (\quad 4, 2, 2, 1, 1, 2)$$
$$\Pi_1 = (\quad 4, 2, 2, 2, 1, 1)$$
$$\Pi'' = (\qquad 1, 1, 1, 0, 1).$$

Clearly Π'' is graphical, so Π is also graphical. The graph so constructed is shown in Fig. 6.2.

The theorem of Erdös and Gallai [EG1] is existential in nature, but its proof uses the same construction.

Theorem 6.2 Let $\Pi = (d_1, d_2, \cdots, d_p)$ be a partition of $2q$ into $p > 1$ parts, $d_1 \geq d_2 \geq \cdots \geq d_p$. Then Π is graphical if and only if for each integer r, $1 \leq r \leq p - 1$,

$$\sum_{i=1}^{r} d_i \leq r(r - 1) + \sum_{i=r+1}^{p} \min \{r, d_i\}. \tag{6.1}$$

Proof. The necessity of these conditions (6.1) is straightforward. Given that Π is a partition of $2q$ belonging to a graph G, the sum of the r largest degrees can be considered in two parts, the first being the contribution to this sum of lines joining the corresponding r points with each other, and the second obtained from lines joining one of these r points with one of the remaining $p - r$ points. These two parts are respectively at most $r(r - 1)$ and $\sum_{i=r+1}^{p} \min \{r, d_i\}$.

The proof of the sufficiency is by induction on p. Clearly the result holds for sequences of two parts. Assume that it holds for sequences of p parts, and let $d_1, d_2, \cdots, d_{p+1}$ be a sequence satisfying the hypotheses of the theorem.

Let m and n be the smallest and largest integers such that

$$d_{m+1} = \cdots = d_{d_1+1} = \cdots = d_n.$$

Form a new sequence of p terms by letting

$$e_i = \begin{cases} d_{i+1} - 1 & \text{for } i = 1 \text{ to } m - 1 \text{ and } n - 1 - (d_1 - m) \text{ to } n - 1, \\ d_{i+1} & \text{otherwise.} \end{cases}$$

If the hypotheses of the theorem hold for the new sequence $e_1, \cdots, e_p$, then by the induction hypothesis, there will be a graph with the numbers e_i as degrees. A graph having the given degree sequence d_i will be formed by adding a new point of degree d_1 adjacent to points of degrees corresponding to those terms e_i which were obtained by subtracting 1 from terms d_{i+1} as above.

Clearly $p > e_1 \geq e_2 \geq \cdots \geq e_p$. Suppose that condition (6.1) does not hold and let h be the least value of r for which it does not. Then

$$\sum_{i=1}^{h} e_i > h(h - 1) + \sum_{i=h+1}^{p} \min \{h, e_i\}. \tag{6.2}$$

But the following inequalities do hold:

$$\sum_{i=1}^{h+1} d_i \leq h(h + 1) + \sum_{i=h+2}^{p+1} \min \{h + 1, d_i\}, \tag{6.3}$$

$$\sum_{i=1}^{h-1} e_i \leq (h - 1)(h - 2) + \sum_{i=h}^{p} \min \{h - 1, e_i\}, \tag{6.4}$$

$$\sum_{i=1}^{h-2} e_i \leq (h - 2)(h - 3) + \sum_{i=h-1}^{p} \min \{h - 2, e_i\}. \tag{6.5}$$

Let s denote the number of values of $i \leq h$ for which $e_i = d_{i+1} - 1$. Then (6.3)–(6.5) when combined with (6.2) yield

$$d_1 + s < 2h + \sum_{i=h+1}^{p} (\min \{h + 1, d_{i+1}\} - \min \{h, e_i\}), \tag{6.6}$$

$$e_h > 2(h - 1) - \min \{h - 1, e_h\} + \sum_{i=h+1}^{p} (\min \{h, e_i\} - \min \{h - 1, e_i\}), \tag{6.7}$$

$$e_{h-1} + e_h > 4h - 6 - \min \{h - 2, e_{h-1}\} - \min \{h - 2, e_h\}$$
$$+ \sum_{i=h+1}^{p} (\min \{h, e_i\} - \min \{h - 2, e_i\}). \tag{6.8}$$

Note that $e_h \geq h$ since otherwise inequality (6.7) gives a contradiction. Let a, b, and c denote the number of values of $i > h$ for which $e_i > h$, $e_i = h$, and $e_i < h$, respectively. Furthermore, let a', b', and c' denote the numbers of these for which $e_i = d_{i+1} - 1$. Then

$$d_1 = s + a' + b' + c'. \tag{6.9}$$

The inequalities (6.6)–(6.8) now become

$$d_1 + s < 2h + a + b' + c', \tag{6.10}$$

$$e_h \geq h + a + b, \tag{6.11}$$

$$e_{h-1} + e_h \geq 2h - 1 + \sum_{i=h+1}^{p} (\min \{h, e_i\} - \min \{h - 2, e_i\}). \tag{6.12}$$

There are now several cases to consider.

CASE 1. $c' = 0$. Since $d_1 \geq e_h$, we have from (6.11),

$$h + a + b \leq d_1.$$

But a combination of (6.9) and (6.10) gives

$$2d_1 < 2h + a + a' + 2b',$$

which is a contradiction.

CASE 2. $c' > 0$ and $d_{h+1} > h$. This means that $d_{i+1} = e_i + 1$ whenever $d_{i+1} > h$. Therefore since $d_{h+1} > h$, $s = h$ and $a = a'$. But the inequalities (6.10) and (6.9) imply that

$$d_1 + h < 2h + a' + b' + c' = d_1 + h,$$

a contradiction.

CASE 3. $c' > 1$ and $d_{h+1} = h$. Under these circumstances, $e_h = h$ and $a = b = 0$, so $d_1 = s + c'$. Furthermore, since $e_h = d_{h+1}$, $e_i = h - 1$ for at least c' values of $i > h$. Hence inequality (6.12) implies

$$e_{h-1} \geq h - 1 + c' > h$$

so that $e_{h-1} = d_h - 1$. Therefore $s = h - 1$, and

$$d_1 = h - 1 + c' \leq e_{h-1} < d_h,$$

a contradiction.

CASE 4. $c' = 1$ and $d_{h+1} = h$. Again, $e_h = h$, $a = b = 0$, and $d_1 = s + c'$. Since $s \leq h - 1$, $d_1 = h$. But this implies $s = 0$ and $d_1 = 1$, so all $d_i = 1$. Thus (6.1) is obviously satisfied, which is a contradiction.

Since $e_h \geq h$ and $d_{h+1} \geq e_h$, we see that d_{h+1} cannot be less than h. Thus all possible cases have been considered and the proof is complete.

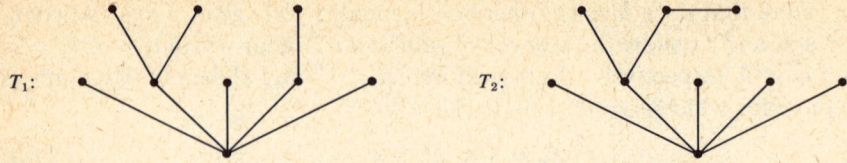

Fig. 6.3. Two trees with the same partition.

Sometimes, it can be determined quite rapidly whether a given partition is graphical and, if it is, the nature of the graphs having this partition may also be discernible. For example, it is easy to give a criterion for a partition to belong to a tree. This result answers a question posed by Ore [O5, p. 62]; it has been found independently many times.

Theorem 6.3 A partition $2q = \Sigma_1^p d_i$ belongs to a tree if and only if each d_i is positive and $q = p - 1$.

As an illustration, consider the partition $16 = 5 + 3 + 2 + 1 + 1 + 1 + 1 + 1 + 1$. Here $d_i > 0$ for each i and $q = 8$ while $p = 9$. Thus Theorem 6.3 assures us that this is the partition of a tree. Two trees to which this partition belongs are shown in Fig. 6.3. But the following graph has the same partition and is not a tree.

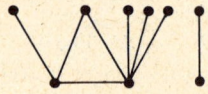

Fig. 6.4.

EXERCISES

6.1 Which of the following partitions are graphical?

   a) $4 + 3 + 3 + 3 + 2 + 2 + 2 + 1$.
   b) $8 + 7 + 6 + 5 + 4 + 3 + 2 + 2 + 1$.
   c) $5 + 5 + 5 + 3 + 3 + 3 + 3 + 3$.
   d) $5 + 4 + 3 + 2 + 1 + 1 + 1 + 1 + 1 + 1 + 1 + 1$.

6.2 Draw all the graphs having the partition $5 + 5 + 3 + 3 + 2 + 2$.

6.3 The partition $16 = 5 + 3 + 2 + 1 + 1 + 1 + 1 + 1 + 1$ belongs to each of the trees in Fig. 6.3. Are there any other trees with this partition?

6.4 Construct all regular graphs with six points.

6.5 Construct all 5 connected cubic graphs with 8 points; all 19 with 10 points.

6.6 There is no graphical partition in which the parts are distinct. Whenever $p \geq 2$, there are exactly two graphs with p points in which just two parts of the partition are equal, and these graphs are complementary.   (Behzad and Chartrand [BC3])

6.7 A graphical partition is *simple* if there is exactly one graph with this partition. Every graphical partition with four parts is simple, and the smallest number of parts in a graphical partition which is not simple is five.

6.8 A partition $(d_1, d_2, \cdots, d_p)$ belongs to a pseudograph (note that a loop contributes 2 to the degree of its point) if and only if $\Sigma \, d_i$ is even.   (Hakimi [H4])

6.9 If a partition of an even integer $2q$ has the form $\Pi = (d_1, d_2, \cdots, d_p)$ with $d_1 \geq d_2 \geq \cdots \geq d_p$, then Π belongs to some multigraph if and only if $q \geq d_1$.

(Hakimi [H4])

*6.10 A partition Π which belongs to some multigraph (see preceding exercise) belongs to exactly one if and only if at least one of the following conditions holds:

1. $p \leq 3$
2. $d_1 = d_2 + \cdots + d_p$
3. $d_1 + 2 = d_2 + \cdots + d_p$ and $d_2 = d_3 = \cdots = d_p$
4. $p = 4$ and $d_3 > d_4 = 1$
5. $d_2 = \cdots = d_p = 1$.   (Senior [S11]; Hakimi [H4])

6.11 Prove or disprove: A tree partition belongs to more than one tree if and only if at least one part is greater than 2, three parts are greater than 1, and if only three, then they are not equal.

6.12 Let $\Pi = (d_1, d_2, \cdots, d_p)$ with $d_1 \geq d_2 \geq \cdots \geq d_p$ and $p \geq 3$ be a graphical partition. Then

Π belongs to some connected graph if and only if $d_p > 0$ and $\Sigma \, d_i \geq 2(p - 1)$.

6.13 A graphical partition Π as in the preceding exercise belongs to some n-line-connected graph with $n \geq 2$ if and only if every $d_i \geq n$.   (Edmonds [E1])

6.14 For any nontrivial graph G and for any partition $p = p_1 + p_2$, there exists a partition $V = V_1 \cup V_2$ such that $|V_i| = p_i$ and $\Delta(\langle V_1 \rangle) + \Delta(\langle V_2 \rangle) \leq \Delta(G)$.

(Lovász [L4])

CHAPTER 7

TRAVERSABILITY

A lie will get you a long way,
but it won't take you home.
ANONYMOUS

One feature of graph theory that has helped to popularize the subject lies in its applications to the area of puzzles and games. Often a puzzle can be converted into a graphical problem: to determine the existence or non-existence of an "eulerian trail" or a "hamiltonian cycle" within a graph. As mentioned in Chapter 1, the concept of an eulerian graph was formulated when Euler studied the problem of the Königsberg bridges. Two characterizations of eulerian graphs are presented. Hamiltonian graphs are studied next and some necessary conditions and some sufficient conditions for graphs to be hamiltonian are given. However, it still remains a challenging unsolved problem to discover an elegant, useful characterization of hamiltonian graphs, rather than only a disguised paraphrase of the definition.

EULERIAN GRAPHS

As we have seen in Chapter 1, Euler's negative solution of the Königsberg Bridge Problem constituted the first publicized discovery of graph theory. The perambulatory problem of crossing bridges can be abstracted to a graphical one: given a graph G, is it possible to find a walk that traverses each line exactly once, goes through all points, and ends at the starting point? A graph for which this is possible is called *eulerian*. Thus, an eulerian graph has an *eulerian trail*, a closed trail containing all points and lines. Clearly, an eulerian graph must be connected.

Theorem 7.1 The following statements are equivalent for a connected graph* G:

(1) G is eulerian.

(2) Every point of G has even degree.

(3) The set of lines of G can be partitioned into cycles.

---

* The theorem clearly holds for multigraphs as well.

Proof. (*1*) *implies* (*2*) Let T be an eulerian trail in G. Each occurrence of a given point in T contributes 2 to the degree of that point, and since each line of G appears exactly once in T, every point must have even degree.

(*2*) *implies* (*3*) Since G is connected and nontrivial, every point has degree at least 2, so G contains a cycle Z. The removal of the lines of Z results in a spanning subgraph G_1 in which every point still has even degree. If G_1 has no lines, then (3) already holds; otherwise, a repetition of the argument applied to G_1 results in a graph G_2 in which again all points are even, etc. When a totally disconnected graph G_n is obtained, we have a partition of the lines of G into n cycles.

(*3*) *implies* (*1*) Let Z_1 be one of the cycles of this partition. If G consists only of this cycle, then G is obviously eulerian. Otherwise, there is another cycle Z_2 with a point v in common with Z_1. The walk beginning at v and consisting of the cycles Z_1 and Z_2 in succession is a closed trail containing the lines of these two cycles. By continuing this process, we can construct a closed trail containing all lines of G; hence G is eulerian.

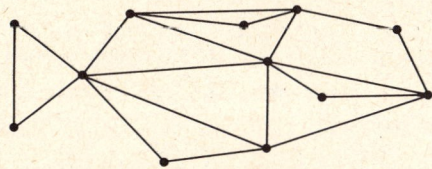

Fig. 7.1. An eulerian graph.

For example, the connected graph of Fig. 7.1 in which every point has even degree has an eulerian trail, and the set of lines can be partitioned into cycles.

By Theorem 7.1 it follows that if a connected graph G has no points of odd degree, then G has a closed trail containing all the points and lines of G. There is an analogous result for connected graphs with some odd points.

Corollary 7.1(a) Let G be a connected graph with exactly $2n$ odd points, $n \geq 1$. Then the set of lines of G can be partitioned into n open trails.

Corollary 7.1(b) Let G be a connected graph with exactly two odd points. Then G has an open trail containing all the points and lines of G (which begins at one of the odd points and ends at the other).

HAMILTONIAN GRAPHS

Sir William Hamilton suggested the class of graphs which bears his name when he asked for the construction of a cycle containing every vertex of a dodecahedron. If a graph G has a spanning cycle Z, then G is called a *hamiltonian graph* and Z a *hamiltonian cycle*. No elegant characterization

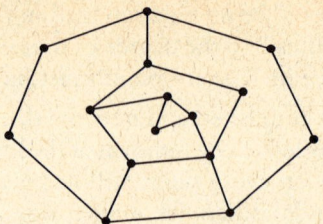

Fig. 7.2. A nonhamiltonian block.

of hamiltonian graphs exists, although several necessary or sufficient conditions are known.

A *theta graph* is a block with two nonadjacent points of degree 3 and all other points of degree 2. Thus a theta graph consists of two points of degree 3 and three disjoint paths joining them, each of length at least 2.

Theorem 7.2 Every hamiltonian graph is 2-connected. Every nonhamiltonian 2-connected graph has a theta subgraph.

It is easy to find a theta subgraph in the nonhamiltonian block of Fig. 7.2.

The next theorem, due to Pósa [P7], gives a sufficient condition for a graph to be hamiltonian. It generalizes earlier results by Ore and Dirac which appear as its corollaries.

Theorem 7.3 Let G have $p \geq 3$ points. If for every n, $1 \leq n < (p - 1)/2$, the number of points of degree not exceeding n is less than n and if, for odd p, the number of points of degree at most $(p - 1)/2$ does not exceed $(p - 1)/2$, then G is hamiltonian.

Proof. Assume the theorem does not hold and let G be a maximal nonhamiltonian graph with p points satisfying the hypothesis of the theorem. It is easy to see that the addition of any line to a graph satisfying the conditions of the theorem results in a graph which also satisfies these conditions. Thus since the addition of any line to G results in a hamiltonian graph, any two nonadjacent points must be joined by a spanning path.

We first show that every point of degree at least $(p - 1)/2$ is adjacent to every point of degree greater than $(p - 1)/2$. Assume (without loss of generality) that $\deg v_1 \geq (p - 1)/2$ and $\deg v_p \geq p/2$, but v_1 and v_p are not adjacent. Then there is a spanning path $v_1 v_2 \cdots v_p$ connecting v_1 and v_p. Let the points adjacent to v_1 be $v_{i_1}, \cdots, v_{i_n}$ where $n = \deg v_1$ and $2 = i_1 < i_2 < \cdots < i_n$. Clearly v_p cannot be adjacent to any point of G of the form $v_{i_j - 1}$, for otherwise there would be a hamiltonian cycle

$$v_1 v_2 \cdots v_{i_j - 1} v_p v_{p-1} \cdots v_{i_j} v_1$$

in G. Now since $n \geq (p - 1)/2$, we have $p/2 \leq \deg v_p \leq p - 1 - n < p/2$ which is impossible, so v_1 and v_p must be adjacent.

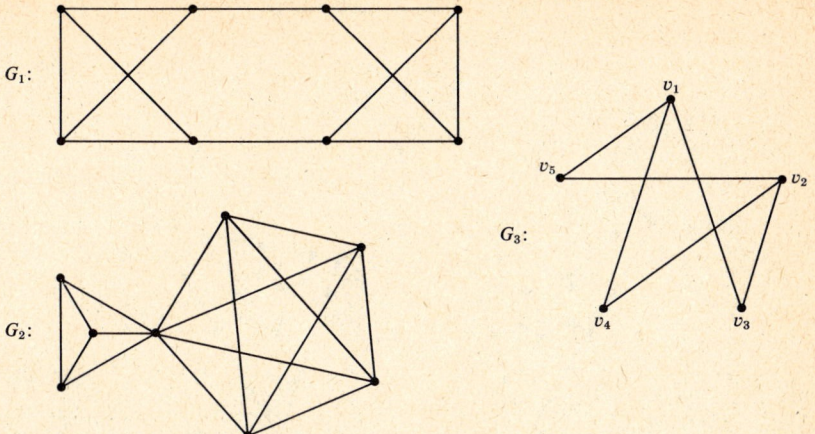

Fig. 7.3. Illustrations for the theorem of Pósa.

It follows that if deg $v \geq p/2$ for all points v, then G is hamiltonian. (This is stated below as Corollary 7.3(b).) For the above argument implies that every pair of points of G are adjacent, so G is complete. But this is a contradiction since K_p is hamiltonian for all $p \geq 3$.

Therefore there is a point v in G with deg $v < p/2$. Let m be the maximum degree among all such points and choose v_1 so that deg $v_1 = m$. By hypothesis the number of points of degree not exceeding m is at most $m < p/2$. Thus there must be more than m points having degree greater than m and hence at least $p/2$. Therefore there is some point, say v_p, of degree at least $p/2$ not adjacent to v_1. Since v_1 and v_p are not adjacent, there is a spanning path $v_1 v_2 \cdots v_p$. As above, we write $v_{i_1}, \cdots, v_{i_m}$ as the points of G adjacent to v_1 and note that v_p cannot be adjacent to any of the m points v_{i_j-1} for $1 \leq j \leq m$. But since v_1 and v_p are not adjacent and v_p has degree at least $p/2$, m must be less than $(p - 1)/2$, by the first part of the proof. Thus, by hypothesis, the number of points of degree at most m is less than m, and so at least one of the m points v_{i_j-1}, say v', must have degree at least $p/2$. We have thus exhibited two nonadjacent points v_p and v', each having degree at least $p/2$, a contradiction which completes the proof.

These sufficient conditions are not necessary. The cubic graph G_1 in Fig. 7.3 is hamiltonian, yet it clearly does not satisfy the conditions of the theorem. However, the theorem is best possible in that no weaker form of it will suffice. For example, choose $p \geq 3$ and $1 \leq n < (p - 1)/2$, and form a graph G_2 with one cutpoint and two blocks, one of which is K_{n+1} and the other K_{p-n}. This graph is not hamiltonian, but it violates the theorem only in that it has exactly n points of degree n. The construction is illustrated in Fig. 7.3 for $p = 8$ and $n = 3$. If we choose $p = 2n + 1$, $n > 1$, and form the graph $G = K_{n,n+1}$, then G is not hamiltonian but violates the theorem only

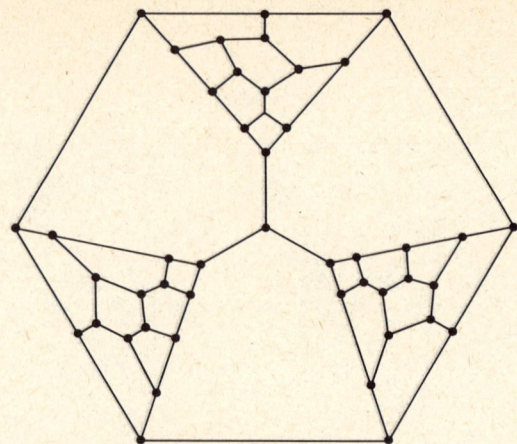

Fig. 7.4. The Tutte graph.

by having $(p - 1)/2 + 1$ points of degree $(p - 1)/2$. The graph $G_3 = K_{2,3}$ of Fig. 7.3 illustrates this construction for $p = 5$.

By specializing Pósa's Theorem, we obtain simpler but less powerful sufficient conditions due to Ore [O3] and Dirac [D6] respectively.

Corollary 7.3(a) If $p \geq 3$ and for every pair u and v of nonadjacent points, $\deg u + \deg v \geq p$, then G is hamiltonian.

Corollary 7.3(b) If for all points v of G, $\deg v \geq p/2$, where $p \geq 3$, then G is hamiltonian.

Actually, the cubic hamiltonian graph G_1 of Fig. 7.3 has four spanning cycles. The smallest cubic hamiltonian graph, K_4, has three spanning cycles. These observations serve to illustrate a theorem of C. A. B. Smith which appears in a paper by Tutte [T6].

Theorem 7.4 Every cubic hamiltonian graph has at least three spanning cycles.

Tait [T1] conjectured that every cubic 3-connected planar graph* contains a spanning cycle. Tutte [T6] settled this in the negative by showing that the 3-connected planar graph with 46 points of Fig. 7.4 is not hamiltonian.

The smallest known nonhamiltonian triply connected cubic planar graph, having 38 points, was constructed independently by J. Lederberg, J. Bosak, and D. Barnette; see Grünbaum [G10, p. 359].

The apparent lack of any relationship between eulerian and hamiltonian graphs is illustrated in Fig. 7.5 where each graph is a block with eight points.

---

* See Chapter 11 for a discussion of planarity. Tait's conjecture, if true, would have settled the Four Color Conjecture.

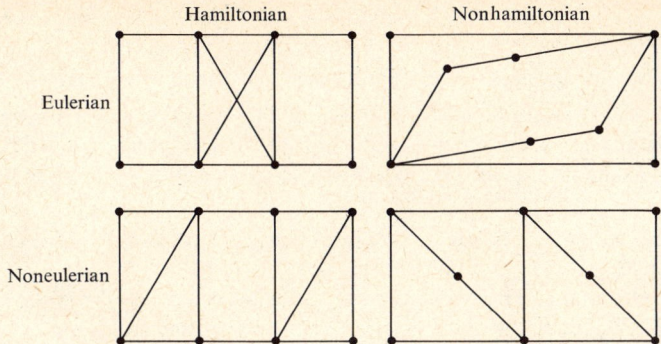

Fig. 7.5. Eulerian and/or hamiltonian graphs.

However, in the next chapter we shall relate eulerian and hamiltonian graphs by way of the "line graph."

Incidentally, M. D. Plummer conjectures that the square of every 2-connected graph is hamiltonian.

EXERCISES

7.1 Find an eulerian trail in the graph G of Fig. 7.1 and a partition of the lines of G into cycles.

7.2 If every block of a connected graph G is eulerian, then G is eulerian, and conversely.

7.3 In Corollary 7.1(a), the partition cannot be done with fewer than n trails. State and prove the converse of Corollary 7.1(b).

7.4 A graph is *arbitrarily traversable from a point* v_0 if the following procedure always results in an eulerian trail: Start at point v_0 by traversing any incident line; on arriving at a point u depart by traversing any incident line not yet used, and continue until no new lines remain.

   a) An eulerian graph is arbitrarily traversable from v_0 if and only if every cycle contains v_0.   (Ore [O2])

   b) If G is arbitrarily traversable from v_0, then v_0 has maximum degree.
                                                        (Bäbler [B1])

   c) If G is arbitrarily traversable from v_0, then either v_0 is the only cutpoint or G has no cutpoints.   (Harary [H17])

7.5 Prove or disprove: If a graph G contains an induced theta subgraph, then G is not hamiltonian.

7.6 a) For any nontrivial connected graph G, every pair of points of G^3 are joined by a spanning path. Hence every line of G^3 is in a hamiltonian cycle, when $p \geq 3$.
                                                        (Karaganis [K2])

   b) If every pair of points of G are joined by a spanning path and $p \geq 4$, then G is 3-connected.

7.7 Give an example of a nonhamiltonian graph with 10 points such that for every pair of nonadjacent points u and v, deg u + deg $v \geq 9$.

7.8 How many spanning cycles are there in the complete bigraphs $K_{3,3}$ and $K_{4,3}$?

7.9 A graph G is called randomly traceable [randomly hamiltonian] if a spanning path [hamiltonian cycle] always results upon starting at any point of G and then successively proceeding to any adjacent point not yet chosen until no new points are available.

    a) A graph G with $p \geq 3$ points is randomly traceable if and only if it is randomly hamiltonian.

    b) A graph G with $p \geq 3$ points is randomly traceable if and only if it is one of the graphs C_p, K_p, or $K_{n,n}$ with $p = 2n$.   (Chartrand and Kronk [CK1])

7.10 Theorem 7.3 can be regarded as giving sufficient conditions for a graph to be 2-connected. This can be generalized to the n-connected case.

Let G be nontrivial and let $1 < n < p$. The following conditions are sufficient for G to be n-connected:

    1. For every k such that $n - 1 \leq k < (p + n - 3)/2$, the number of points of degree not exceeding k does not exceed $k + 1 - n$.

    2. The number of points of degree not exceeding $(p + n - 3)/2$ does not exceed $p - n$.   (Chartrand, Kapoor, and Kronk [CKK1])

7.11 Pósa's theorem can also be generalized in another way.

Let G have $p \geq 3$ and let $0 \leq k \leq p - 2$. If for every integer i with $k + 1 \leq i < (p + k)/2$, the number of points not exceeding i is less than $i - k$, then every path of length k is contained in a hamiltonian cycle.   (Kronk [K13])

7.12 Recall that two labeled graphs are isomorphic if there is a label-preserving isomorphism between them. By an e-graph is meant one in which every point has even degree.

    a) The number of labeled graphs with p points is $2^{p(p-1)/2}$.

    b) The number of labeled e-graphs with p points equals the number of labeled graphs with $p - 1$ points.   (R. W. Robinson)

7.13 If G is a (p, q) graph with $p \geq 3$ and $q \geq (p^2 - 3p + 6)/2$, then G is hamiltonian.   (Ore [O4])

7.14 If for any two nonadjacent points u and v of G, deg u + deg $v \geq p + 1$, then there is a spanning path joining every pair of distinct points.   (Ore [O6])

7.15 If G is a graph with $p \geq 3$ points such that the removal of any set of at most n points results in a hamiltonian graph, then G is $(n + 2)$-connected.

      (Chartrand, Kapoor, Kronk [CKK1])

7.16 Consider the nonhamiltonian graphs G such that every subgraph $G - v$ is hamiltonian. There is exactly one such graph with 10 points and none smaller.

      (Gaudin, Herz, and Rossi [GHR1])

7.17 Do there exist nonhamiltonian graphs with arbitrarily high connectivity?

CHAPTER 8

LINE GRAPHS

A straight line is the shortest distance between two points.

EUCLID

The concept of the line graph of a given graph is so natural that it has been independently discovered by many authors. Of course, each gave it a different name*: Ore [O5] calls it the "interchange graph," Sabidussi [S7] "derivative" and Beineke [B8] "derived graph," Seshu and Reed [SR1] "edge-to-vertex dual," Kasteleyn [K4] "covering graph," and Menon [M10] "adjoint." Various characterizations of line graphs are developed. We also introduce the total graph, first studied by Behzad [B4], which has surprisingly been discovered only once thus far, and hence has no other names. Relationships between line graphs and total graphs are studied, with particular emphasis on eulerian and hamiltonian graphs.

SOME PROPERTIES OF LINE GRAPHS

Consider the set X of lines of a graph G with at least one line as a family of 2-point subsets of $V(G)$. The *line graph* of G, denoted $L(G)$, is the intersection graph $\Omega(X)$. Thus the points of $L(G)$ are the lines of G, with two points of $L(G)$ adjacent whenever the corresponding lines of G are. If $x = uv$ is a line of G, then the degree of x in $L(G)$ is clearly deg u + deg $v - 2$. Two examples of graphs and their line graphs are given in Fig. 8.1. Note that in this figure $G_2 = L(G_1)$, so that $L(G_2) = L(L(G_1))$. We write $L^1(G) = L(G)$, $L^2(G) = L(L(G))$, and in general the *iterated line graph* is $L^n(G) = L(L^{n-1}(G))$.

As an immediate consequence of the definition of $L(G)$, we note that every cutpoint of $L(G)$ is a bridge of G which is not an endline, and conversely.

When defining any class of graphs, it is desirable to know the number of points and lines in each; this is easy to determine for line graphs.

---

* Hoffman [H46] uses "line graph" even though he chooses "edge." Whitney [W11] was the first to discover these graphs but didn't give them a name.

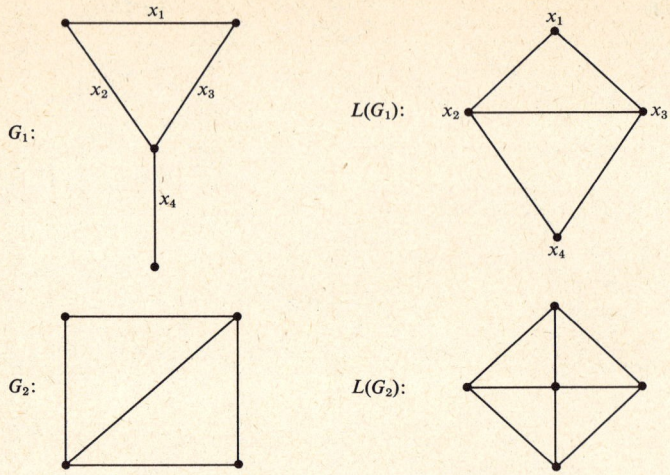

Fig. 8.1. Graphs and their line graphs.

Theorem 8.1 If G is a (p, q) graph whose points have degrees d_i, then $L(G)$ has q points and q_L lines, where

$$q_L = -q + \tfrac{1}{2}\sum d_i^2.$$

Proof. By the definition of line graph, $L(G)$ has q points. The d_i lines incident with a point v_i contribute $\binom{d_i}{2}$ to q_L, so

$$q_L = \sum \binom{d_i}{2} = \tfrac{1}{2}\sum d_i(d_i - 1) = \tfrac{1}{2}\sum d_i^2 - \tfrac{1}{2}\sum d_i = \tfrac{1}{2}\sum d_i^2 - q.$$

The next result can be proved in many different ways, depending on one's whimsy.

Theorem 8.2 A connected graph is isomorphic to its line graph if and only if it is a cycle.

Thus for a (not necessarily connected) graph, $G \cong L(G)$ if and only if G is regular of degree 2.

If G_1 and G_2 are isomorphic, then obviously $L(G_1)$ and $L(G_2)$ are. Whitney [W11] found that the converse almost always holds by displaying the only two different graphs with the same line graph. The proof given here is due to Jung [J3].

Theorem 8.3 Let G and G' be connected graphs with isomorphic line graphs. Then G and G' are isomorphic unless one is K_3 and the other is $K_{1,3}$.

Proof. First note that among the connected graphs with up to four points, the only two different ones with isomorphic line graphs are K_3 and $K_{1,3}$.

Note further that if ϕ is an isomorphism of G onto G', then there is a derived isomorphism ϕ_1 of $L(G)$ onto $L(G')$. The theorem will be demonstrated when the following stronger result is proved.

If G and G' have more than four points, then any isomorphism ϕ_1 of $L(G)$ onto $L(G')$ is derived from exactly one isomorphism of G to G'.

We first show that ϕ_1 is derived from at most one isomorphism. Assume there are two such, ϕ and ψ. We will prove that for any point v of G, $\phi(v) = \psi(v)$. There must exist two lines $x = uv$ and $y = uw$ or vw. If $y = vw$, then the points $\phi(v)$ and $\psi(v)$ are on both lines $\phi_1(x)$ and $\phi_1(y)$, so that since only one point is on both these lines, $\phi(v) = \psi(v)$. By the same argument, when $y = uw$, $\phi(u) = \psi(u)$ so that since the line $\phi_1(x)$ contains the two points $\phi(v)$ and $\phi(u) = \psi(u)$, we again have $\phi(v) = \psi(v)$. Therefore ϕ_1 is derived from at most one isomorphism of G to G'.

We now show the existence of an isomorphism ϕ from which ϕ_1 is derived. The first step is to show that the lines $x_1 = uv_1$, $x_2 = uv_2$, and $x_3 = uv_3$ of a $K_{1,3}$ subgraph of G must go to the lines of a $K_{1,3}$ subgraph of G' under ϕ_1. Let y be another line adjacent with at least one of the x_i, which is adjacent with only one or all three. Such a line y must exist for any graph with $p \geq 5$ and the theorem is trivial for $p < 5$. If the three lines $\phi_1(x_i)$ form a triangle instead of $K_{1,3}$ then $\phi_1(y)$ must be adjacent with precisely two of the three. Therefore, every $K_{1,3}$ must go to a $K_{1,3}$.

Let $S(v)$ denote the set of lines at v. We now show that to each v in G, there is exactly one v' in G' such that $S(v)$ goes to $S(v')$ under ϕ_1. If deg $v \geq 2$, let y_1 and y_2 be lines at v and let v' be the common point of $\phi_1(y_1)$ and $\phi_1(y_2)$. Then for each line x at v, v' is incident with $\phi_1(x)$ and for each line x' at v', v is incident with $\phi_1^{-1}(x')$. If deg $v = 1$, let $x = uv$ be the line at v. Then deg $u \geq 2$ and hence $S(u)$ goes to $S(u')$ and $\phi_1(x) = u'v'$. Since for every line x' at v', the lines $\phi_1^{-1}(x')$ and x must have a common point, u is on $\phi_1^{-1}(x')$ and u' is on x'. That is, $x' = \phi_1(x)$ and deg $v' = 1$. The mapping ϕ is therefore one-to-one from V to V' since $S(u) = S(v)$ only when $u = v$. Now given v' in V', there is an incident line x'. Denote $\phi_1^{-1}(x')$ by uv. Then either $\phi(u) = v'$ or $\phi(v) = v'$ so ϕ is onto.

Finally, we note that for each line $x = uv$ in G, $\phi_1(x) = \phi(u)\phi(v)$ and for each line $x' = u'v'$ in G', $\phi_1^{-1}(x') = \phi^{-1}(u')\phi^{-1}(v')$, so that ϕ is an isomorphism from which ϕ_1 is derived. This completes the proof.

CHARACTERIZATIONS OF LINE GRAPHS

A graph G is a *line graph* if it is isomorphic to the line graph $L(H)$ of some graph H. For example, $K_4 - x$ is a line graph; see Fig. 8.1. On the other hand, we now verify that the star $K_{1,3}$ is not a line graph. Assume $K_{1,3} = L(H)$. Then H has four lines since $K_{1,3}$ has four points, and H must be connected. All the connected graphs with four lines are shown in Fig. 8.2.

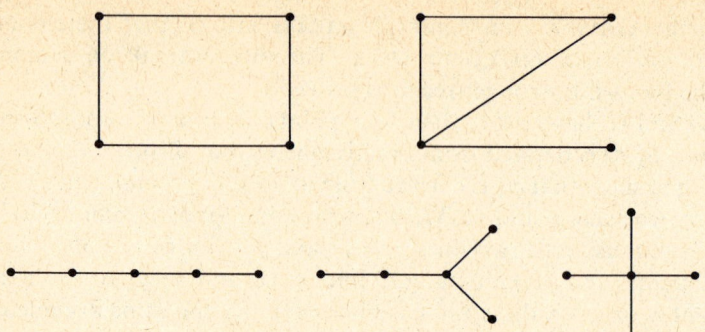

Fig. 8.2. The connected graphs with four lines.

Since $L(C_4) = C_4$ by Theorem 8.3 and $L(K_{1,3} + x) = K_4 - x$ (see Fig. 8.1), it follows that H is one of the three trees. But the line graphs of these trees are the path P_4, the graph $K_3 \cdot K_2$, and K_4, showing that $K_{1,3}$ is not a line graph. We will see that the star $K_{1,3}$ plays an important role in characterizing line graphs. The first characterization of line graphs, statement (2) of the next theorem and due to Krausz [K12], was rather close to the definition. The situation was improved by van Rooij and Wilf [RW1] who were able to describe in (3) a structural criterion for a graph to be a line graph. Finally, Beineke [B8] and N. Robertson (unpublished) displayed exactly those subgraphs which cannot occur in line graphs. Recall that an induced subgraph is one which is maximal on its point set. A triangle T of a graph G is called odd if there is a point of G adjacent to an odd number of its points, and is even otherwise.

Theorem 8.4 The following statements are equivalent:

(1) G is a line graph.

(2) The lines of G can be partitioned into complete subgraphs in such a way that no point lies in more than two of the subgraphs.

(3) G does not have $K_{1,3}$ as an induced subgraph, and if two odd triangles have a common line then the subgraph induced by their points is K_4.

(4) None of the nine graphs of Fig. 8.3 is an induced subgraph of G.

Proof. (1) *implies* (2) Let G be the line graph of H. Without loss of generality we assume that H has no isolated points. Then the lines in the star at each point of H induce a complete subgraph of G, and every line of G lies in exactly one such subgraph. Since each line of H belongs to the stars of exactly two points of H, no point of G is in more than two of these complete subgraphs.

(2) *implies* (1) Given a decomposition of the lines of a graph G into complete subgraphs $S_1, S_2, \cdots, S_n$ satisfying (2), we indicate the construction of a graph H whose line graph is G. The points of H correspond to the set S of

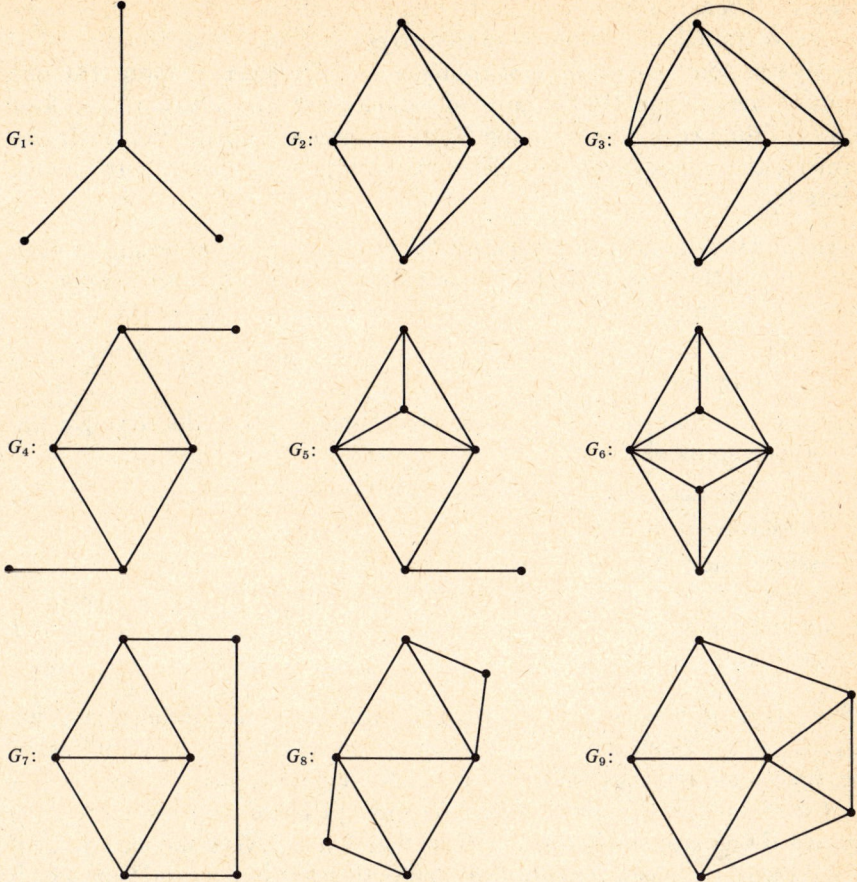

Fig. 8.3. The nine forbidden subgraphs for line graphs.

subgraphs of the decomposition together with the set U of points of G belonging to only one of the subgraphs S_i. Thus $S \cup U$ is the set of points of H and two of these points are adjacent whenever they have a nonempty intersection; that is, H is the intersection graph $\Omega(S \cup U)$.

(2) implies (4) It can be readily verified that none of the nine graphs of Fig. 8.3 can have its set of lines partitioned into complete subgraphs satisfying the given condition. Since every induced subgraph of a line graph must itself be a line graph, the result follows.

(4) implies (3) We show that if G does not satisfy (3), then it has one of the nine forbidden graphs as an induced subgraph. Assume that G has odd triangles abc and abd with c and d not adjacent. There are two cases, depending on whether or not there is a point v adjacent to an odd number of points of both odd triangles.

CASE 1.   There is a point v adjacent to an odd number of points of triangle abc and of triangle abd. Now there are two possibilities: either v is adjacent to exactly one point of each of these triangles or it is adjacent to more than one point of one of them. In the latter situation, v must be adjacent to all four points of the two triangles, giving G_3 as an induced subgraph of G. In the former, either v is adjacent only either to a or b, giving G_1, or to both c and d, giving G_2.

CASE 2.   There is no point adjacent to an odd number of points of both triangles. In this case, let u and v be adjacent to an odd number of points in triangles abc and abd, respectively. There are three subcases to consider:

Case 2.1.   Each of u, v is adjacent to exactly one point of the corresponding triangle.

Case 2.2.   One of u, v is adjacent to all three points of "its" triangle, the other to only one.

Case 2.3.   Each of u, v is adjacent to all three points of the corresponding triangle.

Before these alternatives are considered, we note two facts. If u or v is adjacent to a or b, then it is also adjacent to c or to d, since otherwise G_1 is an induced subgraph. Also, neither u nor v can be adjacent to both c and d since then G_2 or G_3 is induced.

CASE 2.1.   If uc, $vd \in G$ then, depending on whether or not line uv is in G, we have G_4 or G_7 as an induced subgraph. If ub, $vd \in G$ then it follows from the preceding remarks that $ud \in G$ while $vc \notin G$; if $uv \notin G$ then points $\{a, d, u, v\}$ induce G_1, while if $uv \in G$, then $\{a, b, c, d, u, v\}$ induce G_8. If ub, $va \in G$ then necessarily ud, $vc \in G$, so that if $uv \notin G$, G_8 is induced, while if $uv \in G$ then G_2 appears. Finally if ub, $vb \in G$, then again ud, $vc \in G$ from which it follows that either G_9 or G_1 is an induced subgraph of G, depending on whether or not $uv \in G$.

CASE 2.2.   Let ua, ub, $uc \in G$. Clearly if $ud \in G$ then G_3 is induced; thus $ud \notin G$. Now v can be adjacent to d or b. If $vd \in G$, then depending on whether or not $uv \in G$, we find G_2 or G_5 induced. If $vb \in G$ then either G_3 or G_1 is induced, depending on whether or not v is adjacent to both c and u.

CASE 2.3.   If ud, vc, or $uv \in G$, then G_3 is induced. The only other possibility gives G_6.

3. implies 1.   Suppose that G is a graph satisfying the conditions of the statement. We may clearly take G to be connected. Now, exactly one of the following statements must be true:

1. G contains two even triangles with a common line.

2. Whenever two triangles in G have a line in common, one of them is odd.

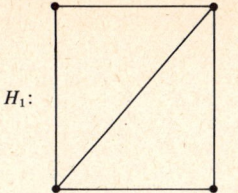 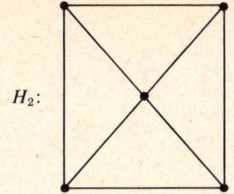 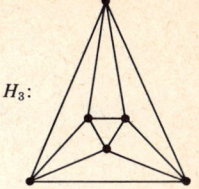

$H_1:$ $H_2:$ $H_3:$

Fig. 8.4. Three line graphs.

It can be shown that if G satisfies the first statement, then it is one of the graphs $H_1 = L(K_{1,3} + x)$, $H_2 = L(H_1)$, or $H_3 = L(K_4)$ displayed in Fig. 8.4. So suppose that G satisfies the second statement. We indicate the method of constructing a graph H such that $G = L(H)$.

Let F_1 be the family of all cliques of G which are not even triangles, where each such clique is considered as a set of points. Let F_2 be the family of points (taken as singletons) of G lying in some clique K in F_1 but not adjacent to any point of $G - K$. Finally, let F_3 be the family of lines (each taken as a set of two points) of G contained in a unique and even triangle. It is not difficult to verify that G is isomorphic to the line graph of the intersection graph $H = \Omega(F_1 \cup F_2 \cup F_3)$. This completes the proof.

This last construction is illustrated in Fig. 8.5, in which the given graph G has families $F_1 = \{\{1, 2, 3, 4\}, \{4, 5, 6\}\}$, $F_2 = \{\{1\}, \{2\}, \{3\}\}$, and $F_3 = \{\{5, 7\}, \{6, 7\}\}$ leading to the intersection graph H; thus $G = L(H)$.

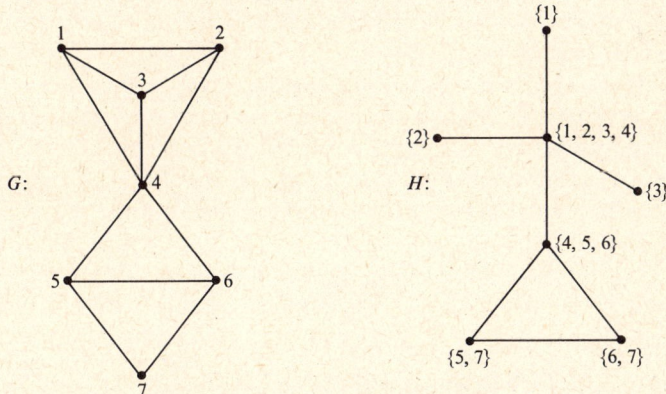

Fig. 8.5. A line graph and its graph.

SPECIAL LINE GRAPHS

In this section, characterizations are presented for line graphs of trees, complete graphs, and complete bigraphs.

The next result, due to G. T. Chartrand, specifies when a graph is the line graph of a tree.

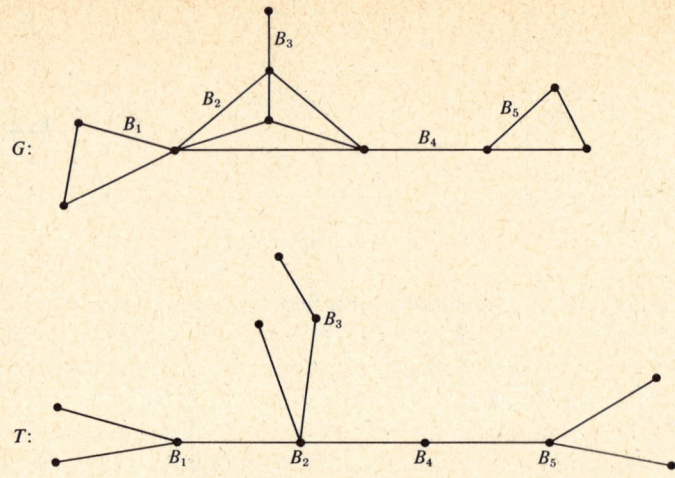

Fig. 8.6. The line-graph G of a tree T.

Theorem 8.5 A graph is the line graph of a tree if and only if it is a connected block graph in which each cutpoint is on exactly two blocks.

Proof. Suppose $G = L(T)$, T some tree. Then G is also $B(T)$ since the lines and blocks of a tree coincide. Each cutpoint x of G corresponds to a bridge uv of T, and is on exactly those two blocks of G which correspond to the stars at u and v. This proves the necessity of the condition.

To see the sufficiency, let G be a block graph in which each cutpoint is on exactly two blocks. Since each block of a block graph is complete, there exists a graph H such that $L(H) = G$ by Theorem 8.4. If $G = K_3$, we can take $H = K_{1,3}$. If G is any other block graph, then we show that H must be a tree. Assume that H is not a tree so that it contains a cycle. If H is itself a cycle, then by Theorem 8.2, $L(H) = H$, but the only cycle which is a block graph is K_3, a case not under consideration. Hence H must properly contain a cycle, thereby implying that H has a cycle Z and a line x adjacent to two lines of Z, but not adjacent to some line y of Z. The points x and y of $L(H)$ lie on a cycle of $L(H)$, and they are not adjacent. This contradicts the condition of Theorem 3.5 that $L(H)$ is a block graph. Hence H is a tree, and the theorem is proved.

In Fig. 8.6, a block graph G is shown in which each cutpoint lies on just two blocks. The tree T of which G is the line graph is constructed by first forming the block graph $B(G)$ and then adding new points for the non-cutpoints of G and the lines joining each block with its noncutpoints.

The line graphs of complete graphs and complete bigraphs are almost always characterized by rather immediate observations involving adjacencies of lines in K_p and $K_{m,n}$. The case of complete graphs was independently settled by Chang [C7] and Hoffman [H43], [H44].

Theorem 8.6 Unless $p = 8$, a graph G is the line graph of K_p if and only if

1. G has $\binom{p}{2}$ points,
2. G is regular of degree $2(p - 2)$,
3. Every two nonadjacent points are mutually adjacent to exactly four points,
4. Every two adjacent points are mutually adjacent to exactly $p - 2$ points.

It is evident that $L(K_p)$ has these four properties. It is not at all obvious that when $p = 8$, there are exactly three exceptional graphs satisfying the conditions.

For complete bigraphs, the corresponding result was found by Moon [M13], and Hoffman [H46].

Theorem 8.7 Unless $m = n = 4$, a graph G is the line graph of $K_{m,n}$ if and only if

1. G has mn points,
2. G is regular of degree $m + n - 2$,
3. Every two nonadjacent points are mutually adjacent to exactly two points,
4. Among the adjacent pairs of points, exactly $n\binom{m}{2}$ pairs are mutually adjacent to exactly $m - 2$ points, and the other $m\binom{n}{2}$ pairs to $n - 2$ points.

There is only one exceptional graph satisfying these conditions. It has 16 points, is not $L(K_{4,4})$, and was found by Shrikhande [S12] when he proved Theorem 8.7 for the case $m = n$.

LINE GRAPHS AND TRAVERSABILITY

We now investigate the relationship of eulerian and hamiltonian graphs with line graphs.

If $x = uv$ is a line of G, and w is not a point of G, then x is *subdivided* when it is replaced by the lines uw and wv. If every line of G is subdivided, the

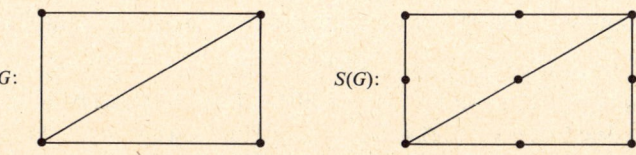

Fig. 8.7. A graph and its subdivision graph.

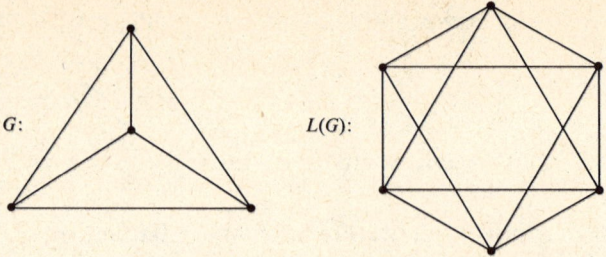

Fig. 8.8. A counterexample.

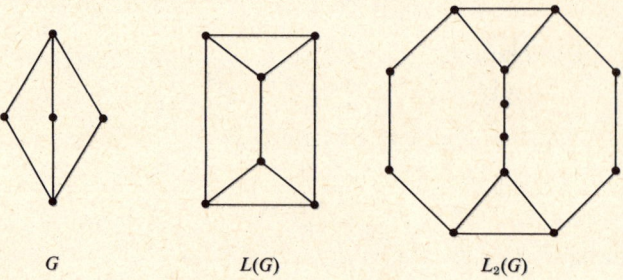

Fig. 8.9. Another counterexample.

resulting graph is the *subdivision graph* $S(G)$; see Fig. 8.7. If we denote by $S_n(G)$ the graph obtained from G by inserting n new points of degree 2 into every line of G, so that $S(G) = S_1(G)$, we can then define a new graph $L_n(G) = L(S_{n-1}(G))$. Note that, in general, $L_n(G) \not\cong L^n(G)$, the nth iterated line graph of G.

Theorem 8.8 If G is eulerian, then $L(G)$ is both eulerian and hamiltonian. If G is hamiltonian, then $L(G)$ is hamiltonian.

It is easy to supply counter-examples to the converses of these statements. For example in Fig. 8.8, $L(G)$ is eulerian and hamiltonian while G is not eulerian; in Fig. 8.9, $L(G)$ is hamiltonian while G is not.

A refinement of the second statement in Theorem 8.8 is provided by the following result of Harary and Nash-Williams [HN1] which follows readily from the preceding theorem and the fact that $L_2(G) = L(S(G))$.

Theorem 8.9 A sufficient condition for $L_2(G)$ to be hamiltonian is that G be hamiltonian and a necessary condition is that $L(G)$ be hamiltonian.

The graphs of Figs. 8.10 and 8.9 show that the first of these conditions is not necessary and the second is not sufficient for $L_2(G)$ to be hamiltonian. We note also (see Fig. 8.11) that $L(G) = L_1(G)$ and $L_2(G)$ may be hamiltonian

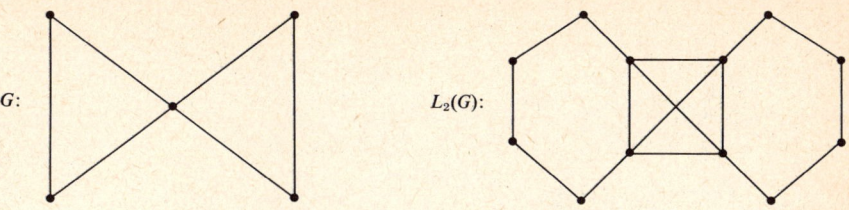

Fig. 8.10. Still another counterexample.

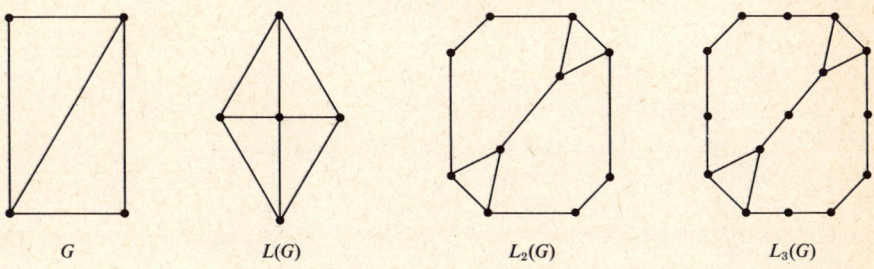

Fig. 8.11. A sequence of graphs $L_n(G)$.

without G being eulerian. However, the next graph $L_3(G)$ in this series provides the link between these two properties.

Theorem 8.10 A graph G is eulerian if and only if $L_3(G)$ is hamiltonian.

For almost every connected graph G, however, nearly all of the graphs $L^n(G)$ are hamiltonian, as shown by Chartrand [C9].

Theorem 8.11 If G is a nontrivial connected graph with p points which is not a path, then $L^n(G)$ is hamiltonian for all $n \geq p - 3$.

An example is given in Fig. 8.12 in which a 6-point graph G, as well as $L(G)$, $L^2(G)$, and the hamiltonian graph $L^3(G)$ are shown.

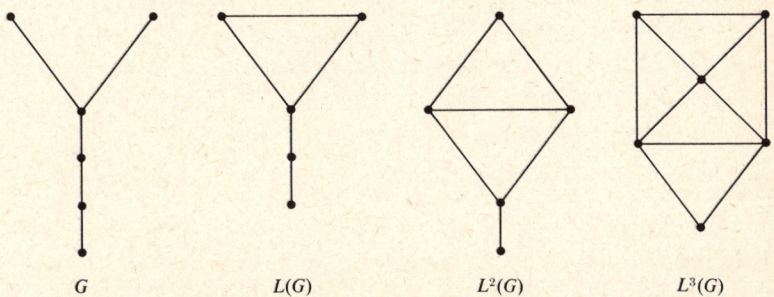

Fig. 8.12. A sequence of iterated line graphs.

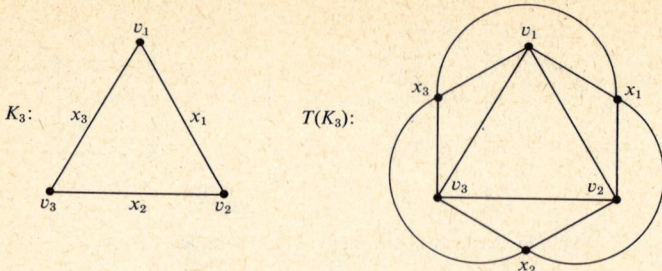

Fig. 8.13. Formation of a total graph.

TOTAL GRAPHS

The points and lines of a graph are called its *elements*. Two elements of a graph are neighbors if they are either incident or adjacent. The *total graph* $T(G)$ has point set $V(G) \cup X(G)$, and two points of $T(G)$ are adjacent whenever they are neighbors in G. Figure 8.13 depicts the formation of the total graph $T(K_3)$. It is easy to see that $T(G)$ always contains both G and $L(G)$ as induced subgraphs.

An alternative characterization of total graphs was given by Behzad [B4].

Theorem 8.12 The total graph $T(G)$ is isomorphic to the square of the subdivision graph $S(G)$.

Corollary 8.12(a) If v is a point of G, then the degree of point v in $T(G)$ is $2 \deg v$. If $x = uv$ is a line of G, then the degree of point x in $T(G)$ is $\deg u + \deg v$.

Corollary 8.12(b) If G is a (p, q) graph whose points have degrees d_i, then the total graph $T(G)$ has $p_T = p + q$ points and $q_T = 2q + \frac{1}{2} \Sigma d_i^2$ lines.

The Ramsey function $r(m, n)$ was defined in Chapter 2, where it was noted that its general determination remains an unsolved problem. Behzad and Radjavi [BR1] defined and solved an analogue of the Ramsey problem, suggested by line graphs. The *line Ramsey number* $r_1(m, n)$ is the smallest positive integer p such that every connected graph with p points contains either n mutually disjoint lines or m mutually adjacent lines, that is, the star $K_{1,m}$. Thus $r_1(m, n)$ is the smallest integer p such that for any graph G with p points, $L(G)$ contains K_m or $\overline{L(G)}$ contains K_n.

Theorem 8.13 For $n > 1$, we always have $r_1(2, n) = 3$. For all other m and n, $r_1(m, n) = (m - 1)(n - 1) + 2$.

Note that it is not always true that $r_1(m, n) = r_1(n, m)$. Furthermore, in contrast with Ramsey numbers, $r_1(m, n)$ is defined only for connected graphs.

EXERCISES

8.1   Under what conditions can the lines of a line graph be partitioned into complete subgraphs so that each point lies in exactly two of these subgraphs?

8.2   Determine the number of triangles in $L(G)$ in terms of the number n of triangles of G and the partition of G.

8.3   Determine a criterion for a connected graph to have a regular line graph.

8.4   A graph G can be reconstructed from the collection of q spanning subgraphs $G - x_j$ if and only if its line graph $L(G)$ satisfies Ulam's Conjecture (p. 12).

(Hemminger [H41])

8.5   If G is n-line-connected, then
1. $L(G)$ is n-connected,
2. $L(G)$ is $(2n - 2)$-line-connected, and
3. $L^2(G)$ is $(2n - 2)$-connected.   (Chartrand and Stewart [CS1])

8.6   a) Construct a connected graph G with $p \geq 4$ such that $L(G)$ is not eulerian but $L^2(G)$ is.
b) There is no connected graph G with $p \geq 5$ such that $L^2(G)$ is not eulerian and $L^3(G)$ is.

8.7   The smallest block whose line graph is not hamiltonian is the theta graph with 8 points in which the distance between the points of degree 3 is 3.   (J. W. Moon)

8.8 $L(G)$ is hamiltonian if and only if there is a closed trail in G which includes at least one point incident with each line of G.

8.9   The graph $L_2(G)$ is hamiltonian if and only if G has a closed spanning trail.

(Harary and Nash-Williams [HN1])

8.10   The following statements are equivalent
(1) $L(G)$ is eulerian.
(2) The degrees of all the points of G are of the same parity and G is connected.
(3) $T(G)$ is eulerian.

8.11 $T(K_p)$ is isomorphic to $L(K_{p+1})$.   (Behzad, Chartrand, and Nordhaus [BCN1])

8.12   Define a family F of subsets of elements of G such that $T(G) = \Omega(F)$.

8.13   a) If G is hamiltonian, so is $T(G)$. If G is eulerian, then $T(G)$ is both eulerian and hamiltonian.
b) The total graph $T(G)$ of every nontrivial connected graph G contains a spanning eulerian subgraph.
c) If a nontrivial graph G contains a spanning eulerian subgraph, then $T(G)$ is hamiltonian.
d) If G is nontrivial and connected, then $T^2(G)$ is hamiltonian.

(Behzad and Chartrand [BC2])

CHAPTER 9

FACTORIZATION

The whole is equal to the sum of its parts.

EUCLID, *Elements*

The whole is greater than the sum of its parts.

MAX WERTHEIMER, *Productive Thinking*

A problem which occurs in varying contexts is to determine whether a given graph can be decomposed into line-disjoint spanning subgraphs possessing a prescribed property. Most frequently, this property is that of regularity of specified degree. In particular, a criterion for the existence in a graph of a spanning regular subgraph of degree 1 was found by Tutte. Some observations are presented concerning the decomposition of complete graphs into spanning subgraphs regular of degree 1 and 2.

The partitioning of the lines of a given graph into spanning forests is also studied and gives rise to an invariant known as "arboricity." A formula for the arboricity of a graph in terms of its subgraphs was derived by Nash-Williams, and explicit constructions for the minimum number of spanning forests in complete graphs and bigraphs have been devised.

1-FACTORIZATION

A *factor* of a graph G is a spanning subgraph of G which is not totally disconnected. We say that G is the *sum** of factors G_i if it is their line-disjoint union, and such a union is called a *factorization* of G. An *n-factor* is regular of degree n. If G is the sum of n-factors, their union is called an *n-factorization* and G itself is *n-factorable*. Unless otherwise stated, the results presented in this chapter appear in or are readily inferred from theorems in König [K10, pp. 155–195], where the topic is treated extensively.

When G has a 1-factor, say G_1, it is clear that p is even and the lines of G_1 are point disjoint. In particular, K_{2n+1} cannot have a 1-factor, but K_{2n} certainly can.

---

* Some call this product; others direct sum.

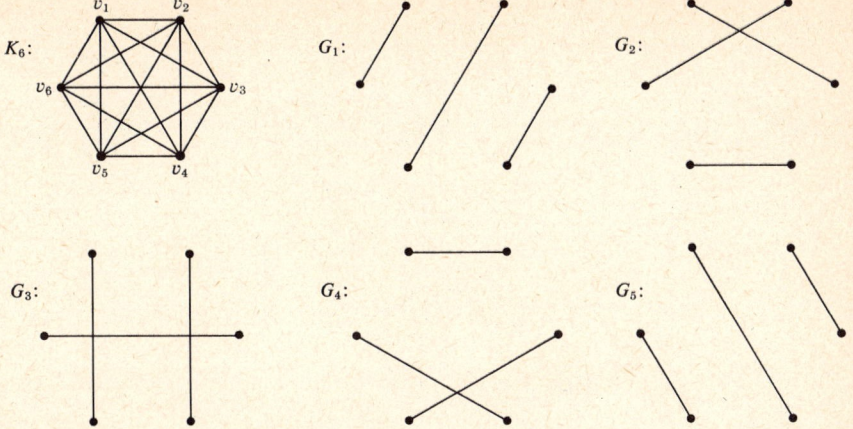

Fig. 9.1. A 1-factorization of K_6.

Theorem 9.1 The complete graph K_{2n} is 1-factorable.

Proof. We need only display a partition of the set X of lines of K_{2n} into $(2n - 1)$ 1-factors. For this purpose we denote the points of G by $v_1, v_2, \cdots, v_{2n}$, and define, for $i = 1, 2, \cdots, 2n - 1$, the sets of lines $X_i = \{v_i v_{2n}\} \cup \{v_{i-j}v_{i+j}; j = 1, 2, \cdots, n - 1\}$, where each of the subscripts $i - j$ and $i + j$ is expressed as one of the numbers $1, 2, \cdots, (2n - 1)$ modulo $(2n - 1)$. The collection $\{X_i\}$ is easily seen to give an appropriate partition of X, and the sum of the subgraphs G_i induced by X_i is a 1-factorization of K_{2n}.

For example, consider the graph K_6 shown in Fig. 9.1. The 1-factorization presented in the proof of the theorem produces the five 1-factors G_i.

Although the complete bigraphs $K_{m,n}$ have no 1-factor if $m \neq n$, the graphs $K_{n,n}$ are 1-factorable, as seen by the next statement.

Theorem 9.2 Every regular bigraph is 1-factorable.

It is not an easy problem to determine whether a given graph is 1-factorable, or, indeed, to establish whether there exists any 1-factor. Beineke and Plummer [BP2] have shown, however, that many graphs cannot have exactly one 1-factor.

Theorem 9.3 If a 2-connected graph has a 1-factor, then it has at least two different 1-factors.

The graph G in Fig. 9.2 is a block with exactly two 1-factors, and they have one common line.

The most significant result on factorization is due to Tutte [T7] and characterizes graphs possessing a 1-factor. In general, this test for a 1-factor is quite inconvenient to apply. The proof given here is based on Gallai [G1].

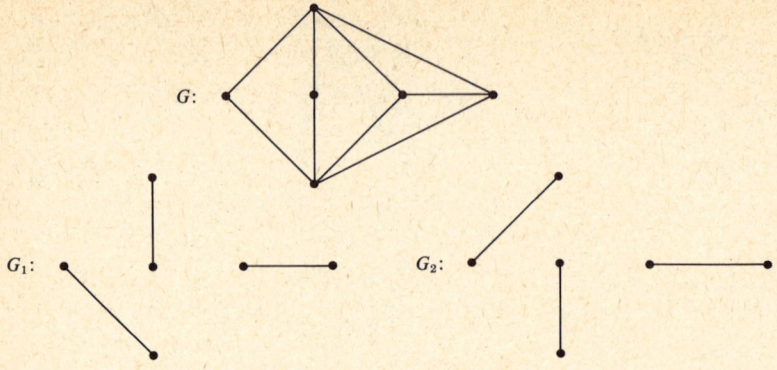

Fig. 9.2. Two 1-factors of a block.

A set of mutually nonadjacent lines is called *independent*. By an odd component of G we mean one with an odd number of points.

Theorem 9.4 A graph G has a 1-factor if and only if p is even and there is no set S of points such that the number of odd components of $G - S$ exceeds $|S|$.

Proof. The easier half of this theorem is its necessity. Let S be any set of points of G and let H be a component of $G - S$. In any 1-factor of G, each point of H must be paired with either another point of H or a point of S. But if H has an odd number of points, then at least one point of H is matched with a point of S. Let k_0 be the number of odd components of $G - S$. If G has a 1-factor then $|S| \geq k_0$, since in a 1-factor each point of S can be matched with at most one point of $G - S$ and therefore can take care of at most one odd component.

In order to prove the sufficiency, assume that G does not have a 1-factor, and let S be a maximum set of independent lines. Let T denote the set of lines not in S, and let u_0 be a point incident only with lines in T. A trail is called alternating if the lines alternately lie in S and T. For each point $v \neq u_0$, call v a 0-point if there are no u_0–v alternating trails; if there is such a trail, call v an S-point if all these trails terminate in a line of S at v, a T-point if each terminates in a line of T at v, and an ST-point if some terminate in each type of line. The following statements are immediate consequences.

Every point adjacent to u_0 is a T- or an ST-point.

No S- or 0-point is adjacent to any S- or ST-point.

No T-point is joined by a line of S to any T- or 0-point.

Therefore, each S-point is joined by a line of S to a T-point. Furthermore, each T-point v is incident with a line of S since otherwise the lines in an alternating u_0–v trail could be switched between S and T to obtain a larger independent set.

Let H be the graph obtained by deleting the T-points. One component of H contains u_0, and any other points in it are ST-points. The other components either consist of an isolated S-point, only ST-points, or only 0-points.

We now show that any component H_1 of H containing ST-points has an odd number of them. Obviously H_1 either contains u_0 or has a point u_1 joined in G to a T-point by a line of S such that some alternating u_0–u_1 trail contains this line and no other points of H_1. If H_1 contains u_0, we take $u_1 = u_0$. The following argument will be used to show that within H_1 every point v other than u_1 is incident with some line of S. This is accomplished by showing that there is an alternating u_1–v trail in H_1 which terminates in a line of S.

The first step in doing this is showing that if there is an alternating u_1–v trail P_1, then there is one which terminates in a line of S. Let P_2 be an alternating u_0–v trail ending in a line of T, and let $u'v'$ be the last line of P_2, if any, which does not lie in H_1. Then u' must be a T-point and $u'v'$ a line in S. Now go along P_1 from u_1 until a point w' of P_2 is reached. Continuing along P_2 in one of the two directions must give an alternating trail. If going to v' results in an alternating path, then the original u_0–u_1 trail P_0 followed by this new path and the line $v'u'$ would be a u_0–u_1 trail terminating in a line of S and u' could not be a T-point. Hence there must be a u_1–v trail terminating in a line of S.

Now we show that there is necessarily a u_1–v alternating trail by assuming there is not. Then there is a point w adjacent to v for which there is a u_1–w alternating trail. If line wv is in S, then the u_1–w alternating trail terminates in a line of T, while if wv is in T, the preceding argument shows there is a u_1–w trail terminating in a line of S. In either case, there is a u_1–v alternating trail.

This shows that the component H_1 has an odd number of points, and that if H_1 does not contain u_0, exactly one of its points is joined to a T-point by a line of S. Hence, with the exception of the component of H containing u_0 and those consisting entirely of 0-points, each is paired with exactly one T-point by a line in S. Since each of these and the component containing u_0 is odd, the theorem is proved.

The graph of Fig. 9.3 has an even number of points but contains no 1-factor, for if the set $S = \{v_1, v_2\}$ is removed from G, four isolated points (and therefore four odd components) remain.

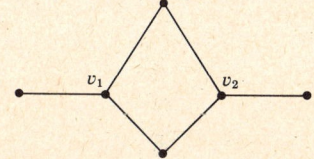

Fig. 9.3. A graph with no 1-factor.

Building up on his criterion for the existence of a 1-factor in a given graph, Tutte [T10] was able to characterize those graphs having a spanning subgraph with prescribed degree sequence, and later, [T11], proved this result as a straightforward consequence of Theorem 9.4. Consider a labeling of G and a function f from V into the nonnegative integers. Let S and T be disjoint subsets of V, let H be a component of $G - (S \cup T)$, and let $q(H, T)$ be the number of lines of G joining a point of H with one in T. Then we may write $k_0(S, T)$ as the number of components H of $G - (S \cup T)$ such that $q(H, T) + \Sigma_{u \in H} f(u)$ is odd.

Theorem 9.5 Let G be a given graph and let f be a function from V into the nonnegative integers. Then G has no spanning subgraph whose degree sequence is prescribed by f if and only if there exist disjoint sets S and T of points such that

$$\sum_{u \in S} f(u) < k_0(S, T) + \sum_{v \in T} [f(v) - d_{G-S}(v)].$$

2-FACTORIZATION

If a graph is 2-factorable, then each factor must be a union of disjoint cycles. If a 2-factor is connected, it is a spanning cycle. We saw that a complete graph is 1-factorable if and only if it has an even number of points. Since a 2-factorable graph must have all points even, the complete graphs K_{2n} are not 2-factorable. The odd complete graphs are 2-factorable, and in fact a stronger statement can be made.

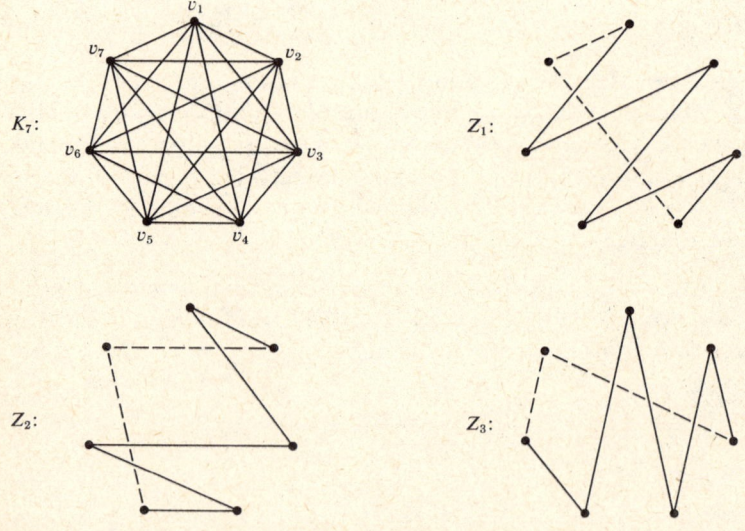

Fig. 9.4. A 2-factorization of K_7.

Theorem 9.6 The graph K_{2n+1} is the sum of n spanning cycles.

Proof. In order to construct n line-disjoint spanning cycles in K_{2n+1}, first label its points $v_1, v_2, \cdots, v_{2n+1}$. Then construct n paths P_i on the points $v_1, v_2, \cdots, v_{2n}$ as follows: $P_i = v_i v_{i-1} v_{i+1} v_{i-2} \cdots v_{i+n-1} v_{i-n}$. Thus the jth point of P_i is v_k, where $k = i + (-1)^{j+1}[j/2]$ and all subscripts are taken as the integers $1, 2, \cdots, 2n \pmod{2n}$. The spanning cycle Z_i is then constructed by joining v_{2n+1} to the endpoints of P_i.

This construction is illustrated in Fig. 9.4 for the graph K_7. The lines of the paths P_i are solid and the two added lines are dashed.

There is a decomposition of K_{2n} which embellishes the result of Theorem 9.1.

Theorem 9.7 The complete graph K_{2n} is the sum of a 1-factor and $n - 1$ spanning cycles.

Of course, every regular graph of degree 1 is itself a 1-factor and every regular graph of degree 2 is a 2-factor. If every component of a regular graph G of degree 2 is an even cycle, then G is also 1-factorable since it can be expressed as the sum of two 1-factors. If a cubic graph contains a 1-factor, it must also have a 2-factor, but there are many cubic graphs which do not have 1-factors.

The graph of Fig. 9.5 has three bridges. Petersen [P3] proved that any cubic graph without a 1-factor must have a bridge.

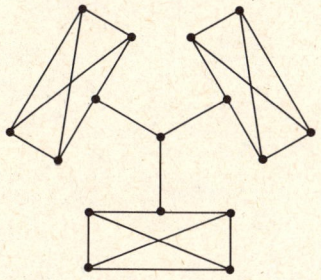
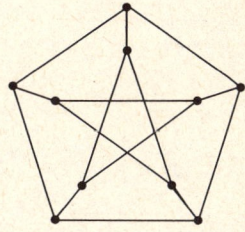

Fig. 9.5. A cubic graph with no 1-factor.   **Fig. 9.6.** The Petersen graph.

Theorem 9.8 Every bridgeless cubic graph is the sum of a 1-factor and a 2-factor.

Petersen showed that this result could not be strengthened by exhibiting a bridgeless cubic graph which is not the sum of three 1-factors. This well-known graph, shown in Fig. 9.6, is called the *Petersen graph*. By Theorem 9.8, it is the sum of a 1-factor and a 2-factor. The pentagon and pentagram together constitute a 2-factor while the five lines joining the pentagon with the pentagram form a 1-factor.

A criterion for the decomposability of a graph into 2-factors was also obtained by Petersen [P3].

Theorem 9.9 A graph is 2-factorable if and only if it is regular of even degree. even degree.

ARBORICITY

In the only type of factorization considered thus far, each factor has been an n-factor. Several other kinds of factorizations have been investigated and we discuss one now and others in Chapter 11. Any graph G can be expressed as a sum of spanning forests, simply by letting each factor contain only one of the q lines of G. A natural problem is to determine the minimum number of line-disjoint spanning forests into which G can be decomposed. This number is called the *arboricity* of G and is denoted by $\Upsilon(G)$. For example, $\Upsilon(K_4) = 2$ and $\Upsilon(K_5) = 3$; minimal decompositions of these graphs into spanning forests are shown in Fig. 9.7.

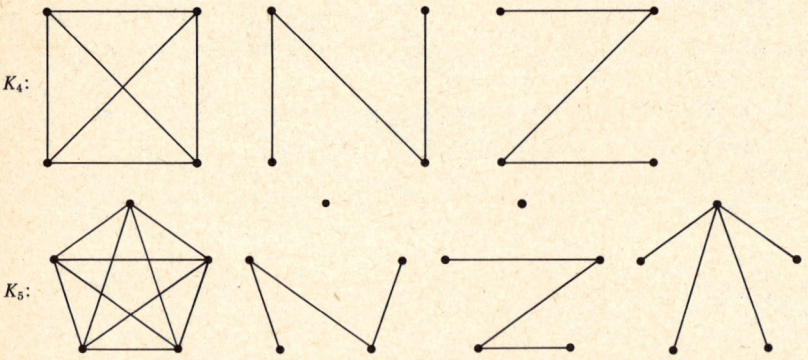

Fig. 9.7. Minimal decompositions into spanning forests.

A formula discovered by Nash-Williams [N2] gives the arboricity of any graph.

Theorem 9.10 Let G be a nontrivial (p, q) graph and let q_n be the maximum number of lines in any subgraph of G having n points. Then

$$\Upsilon(G) = \max_n \left\{ \frac{q_n}{n - 1} \right\}.$$

The fact that $\Upsilon(G) \geq \max_n \{q_n/(n - 1)\}$ can be shown as follows. Since G has p points, the maximum number of lines in any spanning forest is $p - 1$. Hence, the minimum possible number of spanning forests required to fill G, which by definition is $\Upsilon(G)$, is at least $q/(p-1)$. But the arboricity of G is an integer, so $\Upsilon(G) \geq \{q/(p-1)\}$. The desired inequality now follows from the fact that for any subgraph H of G, $\Upsilon(G) \geq \Upsilon(H)$.

Among all subgraphs H with $n \leq p$ points, max $\Upsilon(H)$ will occur in those induced subgraphs containing the greatest number of lines. Thus if H is a subgraph of G, $\Upsilon(H)$ can be greater than $\{q/(p-1)\}$. The (10, 15) graph in Fig. 9.8 illustrates this observation. Taking $n = 5$ and $q_n = 10$ (for $H = K_5$), we have

$$\Upsilon(H) \geq \left\{\frac{q_n}{n-1}\right\} = 3 > 2 = \left\{\frac{q}{p-1}\right\}.$$

For K_p, the maximum value of $q_n/(n-1)$ clearly occurs for $n = p$ so that $\Upsilon(K_p) = \{p/2\}$. Similarly, for the complete bigraph $K_{r,s}$, $\{q_n/(n-1)\}$ assumes its maximum value when $n = p = r + s$.

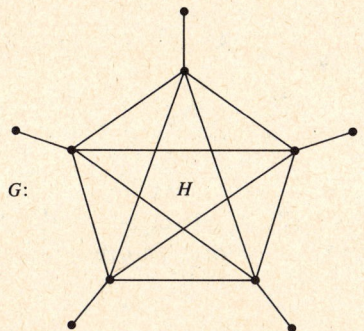

G: H

Fig. 9.8. A graph G with a dense subgraph H.

Corollary 9.10(a) The arboricities of the complete graphs and bigraphs are

$$\Upsilon(K_p) = \left\{\frac{p}{2}\right\} \quad \text{and} \quad \Upsilon(K_{r,s}) = \left\{\frac{rs}{r+s-1}\right\}.$$

Although Nash-Williams' formula gives the minimum number of spanning forests into which an arbitrary graph can be factored, his proof does not display a specific decomposition. Beineke [B5] accomplished this for complete graphs and bigraphs, the former of which we present here. For $p = 2n$, K_p can actually be decomposed into n spanning paths. Labeling the points $v_1, v_2, \cdots, v_{2n}$, we consider the same n paths

$$P_i = v_i \, v_{i-1} \, v_{i+1} \, v_{i-2} \, v_{i+2} \cdots v_{i+n-1} \, v_{i-n},$$

as in proof of Theorem 9.6. For $p = 2n + 1$, the arboricity of K_p is $n + 1$ by Corollary 9.10(a). A decomposition is obtained by taking the paths just described, adding an extra point labeled v_{2n+1} to each, and then constructing

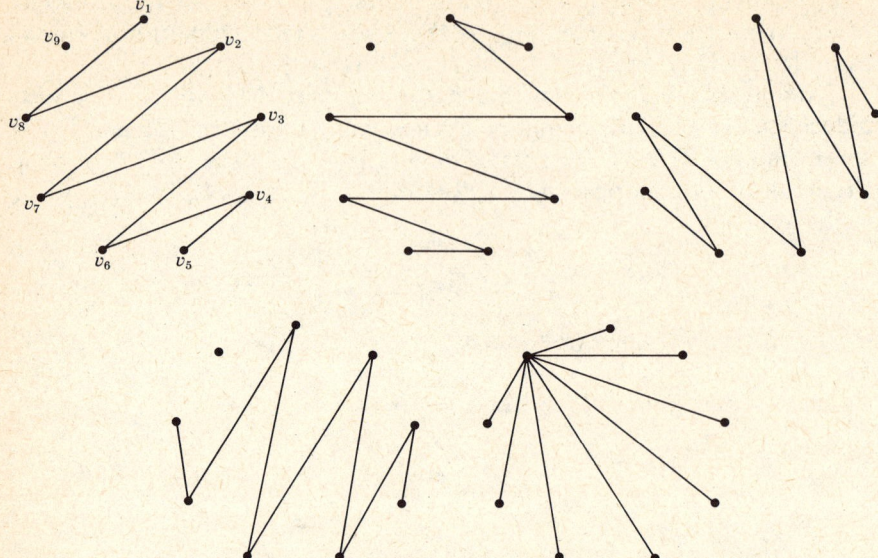

Fig. 9.9. A minimal decomposition of K_9 into spanning forests.

a star by joining v_{2n+1} to the other $2n$ points. The construction for $p = 9$ is shown in Fig. 9.9. It is easily seen to consist of the star at one of the points of K_9 together with spanning subforests corresponding to the four spanning paths of K_8 indicated above.

EXERCISES

9.1 The graph K_4 has a unique 1-factorization. Find the number of 1-factorizations of $K_{3,3}$ and of K_6.

9.2 Display a 1-factorization for K_8.

9.3 The number of 1-factors in K_{2n} is $(2n)!/(2^n n!)$.

9.4 K_{6n-2} has a 3-factorization.

9.5 For $n \geq 1$, K_{4n+1} is 4-factorable.

9.6 Use Tutte's Theorem 9.4 to show that the graph of Fig. 9.5 has no 1-factor.

9.7 If an n-connected graph G with p even is regular of degree n, then G has a 1-factor.

(Tutte [T7])

9.8 Prove or disprove: Let G be a graph with a 1-factor F. A line of G is in more than one 1-factor if and only if it lies on a cycle whose lines are alternately in F.

(Beineke and Plummer [BP2])

9.9 Express K_9 as the sum of four spanning cycles.

9.10 Is the Petersen graph hamiltonian?

*9.11 Corresponding to any two integers $d \geq 3$ and $g \geq 3$, there exists a graph G with the following properties:

1. G is regular of degree d.
2. G has girth g.
3. G is hamiltonian.
4. The cycles of length g are line-disjoint and constitute a 2-factor of G.
5. G is the sum of this 2-factor and $(d - 2)$ 1-factors.   (Sachs [S9])

9.12 Display a minimal decomposition of $K_{4,4}$ into spanning forests.

9.13 Find the smallest connected (p, q) graph G such that

$$\max_r \{q_r/(r - 1)\} > \{q/(p - 1)\},$$

where q_r is the maximum number of lines in any induced subgraph of G with r points.

CHAPTER 10

COVERINGS

> Through any point not on a given line, there passes
> a unique line having no points in common with the given line.
>
> EUCLID
>
> Through any point not on a given line, there passes
> no line having no points in common with the given line.
>
> RIEMANN
>
> Through any point not on a given line, there pass
> more than one line having no points in common with the given line.
>
> BOLYAI

It is natural to say that a line $x = uv$ of G covers the points u and v. Similarly, we may consider each point as covering all lines incident with it. From this viewpoint, one defines two invariants of G: the minimum number of points (lines) which cover all the lines (points). Two related invariants are the maximum number of nonadjacent points and lines. These four numbers associated with any graph satisfy several relations and also suggest the study of special points and lines which are critical for covering purposes. These concepts lead naturally to two special subgraphs of G called the line-core and point-core. Criteria for the existence of such subgraphs are established in terms of covering properties of the graph.

COVERINGS AND INDEPENDENCE

A point and a line are said to *cover* each other if they are incident. A set of points which covers all the lines of a graph G is called a *point cover* for G, while a set of lines which covers all the points is a *line cover*. The smallest number of points in any point cover for G is called its *point covering number* and is denoted by $\alpha_0(G)$ or α_0. Similarly, $\alpha_1(G)$ or α_1 is the smallest number of lines in any line cover of G and is called its *line covering number*. For example, $\alpha_0(K_p) = p - 1$ and $\alpha_1(K_p) = [(p + 1)/2]$. A point cover (line cover) is called *minimum* if it contains α_0 (respectively α_1) elements. Observe that a

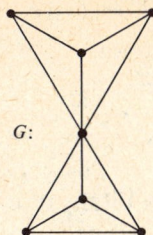

G:

Fig. 10.1. The graph $K_4 \cdot K_4$.

point cover may be minimal without being minimum; such a set of points is given by the 6 noncutpoints in Fig. 10.1. The same holds for line covers; the 6 lines incident with the cutpoint serve.

A set of points in G is *independent* if no two of them are adjacent. The largest number of points in such a set is called the *point independence number* of G and is denoted by $\beta_0(G)$ or β_0. Analogously, an *independent set of lines* of G has no two of its lines adjacent and the maximum cardinality of such a set is the *line independence number* $\beta_1(G)$ or β_1. For the complete graph, $\beta_0(K_p) = 1$ and $\beta_1(K_p) = [p/2]$. Obviously $\beta_1(G) = p/2$ if and only if G has a 1-factor. The numbers just defined are $\beta_0(G) = 2$ and $\beta_1(G) = 3$ for the graph G of Fig. 10.1.

For this graph as well as for K_p, $\alpha_0 + \beta_0 = \alpha_1 + \beta_1 = p$. Gallai [G2] proved that this identity always holds.

Theorem 10.1 For any nontrivial connected graph G,

$$\alpha_0 + \beta_0 = p = \alpha_1 + \beta_1.$$

Proof. Let M_0 be any maximum independent set of points, so that $|M_0| = \beta_0$. Since no line joins two points of M_0, the remaining set of $p - \beta_0$ points constitutes a point cover for G so that $\alpha_0 \leq p - \beta_0$. On the other hand, if N_0 is a minimum point cover for G, then no line can join any two of the remaining $p - \alpha_0$ points of G, so the set $V - N_0$ is independent. Hence, $\beta_0 \geq p - \alpha_0$, proving the first equation.

To obtain the second equality, we begin with an independent set M_1 of β_1 lines. A line cover Y is then produced by taking the union of M_1 and a set of lines, one incident line for each point of G not covered by any line in M_1. Since $|M_1| + |Y| \leq p$ and $|Y| \geq \alpha_1$, it follows that $\alpha_1 + \beta_1 \leq p$. In order to show the inequality in the other direction, let us consider a minimum line cover N_1 of G. Clearly, N_1 cannot contain a line both of whose endpoints are incident with lines also in N_1. This implies that N_1 is the sum of stars of G (considered as sets of lines). If one line is selected from each of these stars, we obtain an independent set W of lines. Now, $|N_1| + |W| = p$ and $|W| \leq \beta_1$; thus, $\alpha_1 + \beta_1 \geq p$, completing the proof of the theorem.

Hedetniemi [H39] noticed that the proof of the first equation in Theorem 10.1,

$$\alpha_0 + \beta_0 = p,$$

applies in a more general setting. A property P of a graph G is *hereditary* if every subgraph of G also has this property. Examples of hereditary properties include a graph being totally disconnected, acyclic, and bipartite. A set S of points of G is called a P-set if the induced subgraph $\langle S \rangle$ has property P; it is called a $\bar{P}$-set if every subgraph of G without property P contains a point of S. Let $\beta_0(P)$ be the maximum cardinality of a P-set of G and let $\alpha_0(P)$ be the minimum number of points of a $\bar{P}$-set. Then the proof of the next statement is obtained at once from that of Theorem 10.1.

Corollary 10.1(a)   If P is an hereditary property of G, then $\alpha_0(P) + \beta_0(P) = p$.

A collection of independent lines of a graph G is sometimes called a *matching* of G since it establishes a pairing of the points incident to them. For this reason, a set of β_1 independent lines in G is called a *maximum matching* of G. If G is bipartite, then more can be said. The next theorem due to König [K9] is intimately related to his Theorem 5.18 on systems of distinct representatives stated in matrix form, in fact it is the same result.

Theorem 10.2   If G is bipartite, then the number of lines in a maximum matching equals the point covering number, that is, $\beta_1 = \alpha_0$.

The problem of finding a maximum matching, the so-called matching problem, is closely related to that of finding a minimum point cover.

Let $M \subset X(G)$ be a matching of G. In an alternating M-trail, exactly one of any two consecutive lines is in M. An augmenting M-trail is an alternating M-trail whose endpoints are not incident with any line of M. Such a trail must be a path or cycle because M is a matching. If G has no augmenting M-trail, then matching M is *unaugmentable*. Clearly every maximum matching is unaugmentable; the converse is due to Berge [B10] and the proof given below appears in Norman and Rabin [NR1].

Theorem 10.3   Every unaugmentable matching is maximum.

Proof. Let M be unaugmentable and choose a maximum matching M' for which $|M - M'|$, the number of lines which are in M but not in M', is minimum. If this number is zero then $M = M'$. Otherwise, construct a trail W of maximum length whose lines alternate in $M - M'$ and M'. Since M' is unaugmentable, trail W cannot begin and end with lines of $M - M'$ and has equally many lines in $M - M'$ and in M'. Now we form a maximum matching N from M' by replacing those lines of W which are in M' by the lines of W in $M - M'$. Then $|M - N| < |M - M'|$, contradicting the choice of M' and completing the proof.

Norman and Rabin [NR1] developed an algorithm, based on the next theorem, for finding all minimum line covers in a given graph. Let Y be a line cover of G. An alternating Y-walk is a Y-reducing walk if its endlines are in Y and its endpoints are incident to lines of Y which are not endlines of the walk. Obviously every minimum line cover has no reducing walk.

Theorem 10.4 If Y is a line cover of G such that there is no Y-reducing walk, then Y is a minimum line cover.

The cover invariants α_0 and α_1 of G refer to the number of points needed to cover all the lines and vice versa. We may also regard each point as covering itself and two points as covering each other if they are adjacent, and similarly for lines. Then other invariants suggest themselves.

Let α_{00} be the minimum number of points needed to cover V, and let α'_{00} be the minimum number* of independent points which cover V. Then both these numbers are defined for any graph. Let α_{11} and α'_{11} have similar meanings for the covering of lines by lines. The relationships among these invariants were determined by Gupta [G11].

Theorem 10.5 For any graph G,

$$\alpha_{00} \leq \alpha'_{00} \quad \text{and} \quad \alpha_{11} = \alpha'_{11}.$$

CRITICAL POINTS AND LINES

Obviously, if H is a subgraph of G, then $\alpha_0(H) \leq \alpha_0(G)$. In particular, this inequality holds when $H = G - v$ or $H = G - x$ for any point v or line x. If $\alpha_0(G - v) < \alpha_0(G)$, then v is called a *critical*† *point*; if $\alpha_0(G - x) < \alpha_0(G)$, then x is a *critical line* of G. Clearly, if v and x are critical, it follows that $\alpha_0(G - v) = \alpha_0(G - x) = \alpha_0 - 1$. Critical points are easily characterized.

Theorem 10.6 A point v is critical in a graph G if and only if some minimum point cover contains v.

Proof. If M is a minimum point cover for G which contains v, then $M - \{v\}$ covers $G - v$; hence, $\alpha_0(G - v) \leq |M - \{v\}| = |M| - 1 = \alpha_0(G) - 1$ so that v is critical in G.

Let v be a critical point of G and consider a minimum point cover M' for $G - v$. The set $M' \cup \{v\}$ is a point cover for G, and since it contains one more element than M', it is minimum.

If the removal of a line $x = uv$ from G decreases the point covering number, then the removal of u or v must also result in a graph with smaller point covering number. Thus, if a line is critical both its endpoints are

---

* Berge [B12] calls α_{00} the "external stability number" and β_0 the "internal stability number."
† In this chapter, "critical" refers to covering; in Chapter 12, the same word will involve coloring. The meanings should be clear by context.

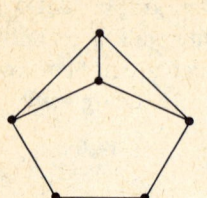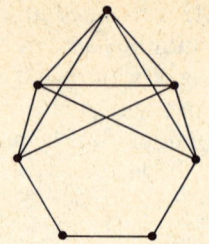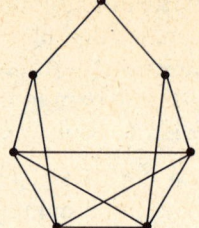

Fig. 10.2. Line-critical graphs.

critical. If a graph has critical points, it need not have critical lines; for example, every point of C_4 is critical but no line is.

A graph in which every point is critical is called *point-critical* while one having all lines critical is *line-critical*. Thus a graph G is point-critical if and only if each point of G lies in some minimum point cover for G. From our previous remarks, every line-critical graph without isolated points is point-critical. Among the line-critical graphs are the complete graphs, the cycles of odd length, and the graphs of Fig. 10.2.

A constructive criterion for line-critical graphs is not known at present; however, the first two corollaries to the following theorem of Beineke, Harary, and Plummer [BHP1] place some rather stringent conditions on such graphs.

Theorem 10.7 Any two adjacent critical lines of a graph lie on an odd cycle.

Corollary 10.7(a) Every connected line-critical graph is a block in which any two adjacent lines lie on an odd cycle.

Theorem 10.7 was derived by generalizing the next result due to Dulmage and Mendelsohn [DM1].

Corollary 10.7(b) Any two critical lines of a bipartite graph are independent.

LINE-CORE AND POINT-CORE

The *line-core** $C_1(G)$ of a graph G is the subgraph of G induced by the union of all independent sets Y of lines (if any) such that $|Y| = \alpha_0(G)$. This concept was introduced by Dulmage and Mendelsohn [DM1], who made it an integral part of their theory of decomposition for bipartite graphs. It is not always the case that a graph has a line-core, though by Theorem 10.2, every bipartite graph which is not totally disconnected has one. As an example of a graph with no line-core, consider an odd cycle C_p. Here we find that $\alpha_0(C_p) = (p + 1)/2$ but that $\beta_1(C_p) = (p - 1)/2$, so C_p has no line-core.

---

* Called "core" in [DM1] and [HP19].

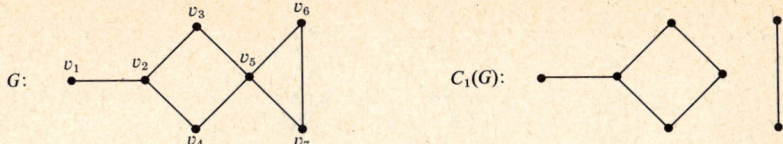

Fig. 10.3. A graph and its line-core.

Harary and Plummer [HP19] developed a criterion for a graph to have a line-core. A minimum point cover M for a graph G with point set V is said to be external if for each subset M' of M, $|M'| \le |U(M')|$, where $U(M')$ is the set of all points of $V - M$ which are adjacent to a point of M'.

Theorem 10.8 The following are equivalent for any graph G:

(1) G has a line-core.

(2) G has an external minimum point cover.

(3) Every minimum point cover for G is external.

As an example, consider the graph G of Fig. 10.3. This graph has two minimum point covers: $M_1 = \{v_2, v_5, v_6\}$ and $M_2 = \{v_2, v_5, v_7\}$. Let us concentrate on M_1. If $M'_1 = M_1$, then $U(M'_1) = \{v_1, v_3, v_4, v_7\}$. For $M''_1 = \{v_5, v_6\}$, $U(M''_1) = \{v_3, v_4, v_7\}$. We observe that $|M'_1| \le |U(M'_1)|$ and $|M''_1| \le |U(M''_1)|$, a fact which is true for every subset of M_1; hence, by definition, M_1 is external. Obviously, M_2 is also external.

On the other hand, there are graphs which are equal to their line-core. This family of graphs is characterized in the next theorem, given in [HP19]. Following the terminology of Dulmage and Mendelsohn [DM1], we consider a bigraph G whose point set V is the disjoint union $S \cup T$. We say that G is *semi-irreducible* if G has exactly one minimum point cover M and either $M \cap S$ or $M \cap T$ is empty. Next, G is *irreducible* if it has exactly two minimum point covers M_1 and M_2 and either $M_1 \cap S = \phi$ and $M_2 \cap T = \phi$ or $M_1 \cap T = \phi$ and $M_2 \cap S = \phi$. Finally, G is *reducible* if it is neither irreducible nor semi-irreducible.

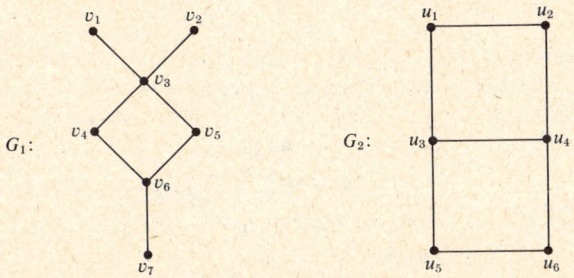

Fig. 10.4. A semi-irreducible and an irreducible graph.

Theorem 10.9 A graph G and its line-core $C_1(G)$ are equal if and only if G is bipartite and not reducible.

Consider the bigraphs G_1 and G_2 of Fig. 10.4. In G_1, let $S_1 = \{v_3, v_6\}$ and $T_1 = \{v_1, v_2, v_4, v_5, v_7\}$. The bigraph G_1 has the unique minimum point cover $M_1 = \{v_3, v_6\}$, and since $M_1 \cap T_1 = \phi$, G_1 is semi-irreducible and hence equals its line-core. In G_2, set $S_2 = \{u_1, u_4, u_5\}$ and $T_2 = \{u_2, u_3, u_6\}$. There are two minimum point covers, namely, $M_2 = \{u_1, u_4, u_5\}$ and $N_2 = \{u_2, u_3, u_6\}$. However, $M_2 \cap T_2 = \phi$ and $N_2 \cap S_2 = \phi$; therefore, G_2 is irreducible and also equals its line-core.

EXERCISES

10.1 Prove or disprove: Every point cover of a graph G contains a minimum point cover.

10.2 Prove or disprove: Every independent set of lines is contained in a maximum independent set of lines.

10.3 For any graph G, $\alpha_0(G) \geq \beta_1(G)$ and $\alpha_1(G) \geq \beta_0(G)$.

10.4 Find a necessary and sufficient condition that $\alpha_1(G) = \beta_1(G)$.

10.5 If G has a closed trail containing a point cover, then $L(G)$ is hamiltonian.

10.6 For any graph G, $\alpha_0(G) \geq \delta(G)$.

10.7 If G is a bigraph then $q \leq \alpha_0 \beta_0$, with equality holding only for complete bigraphs.

10.8 If G is a complete n-partite graph, then

    a) $\alpha_0 = \delta = \kappa = \lambda$.

    b) G is hamiltonian if and only if $p \leq 2\alpha_0$.

    c) If G is not hamiltonian, then its circumference $c = 2\alpha_0$ and G has a unique minimum point cover.   (M.D. Plummer)

    d) $\beta_1 = \min \{\delta, [p/2]\}$.   (Chartrand, Geller, Hedetniemi [CGH2])

10.9 a) Let β_κ be the maximum number of points in a set $S \subset V(G)$ such that $\langle S \rangle$ is disconnected. Then $\kappa = p - \beta_\kappa$.

    b) Defining β_λ analogously, $\lambda = q - \beta_\lambda$.   (Hedetniemi [H39])

10.10 Calculate:

    a) $\alpha_{11}(K_p)$,   b) $\alpha_{00}(K_{m,n})$,   c) $\alpha_{11}(K_{m,n})$.

10.11 The "chess-queen graph" has the 64 squares of a chess board as its points, two of which are adjacent whenever one can be reached from the other by a single move of a queen; the chess-knight, chess-bishop, and chess-rook graphs are defined similarly. What is the number α_{00} for each of these four graphs?

    (Solutions are displayed in Berge [B12, pp. 41–42])

10.12 Some relationships among α_0, α_{00}, and α'_{00} are as follows:

    a) $\alpha_{00} \leq \alpha_0$ if there are no isolated points.

    b) For some graphs, $\alpha_0 < \alpha'_{00}$.

    c) For some graphs, $\alpha'_{00} < \alpha_0$.

    d) For some graphs, $\alpha_{00} < \alpha'_{00}$.

10.13   Prove or disprove: A line x is critical in a graph G if and only if there is a minimum line cover containing x.

10.14   Prove or disprove: Every 2-connected line-critical graph is hamiltonian.

10.15   The converse of Corollary 10.7(a) does not hold.   Construct a block which is not line-critical in which any two adjacent lines lie on an odd cycle.

10.16   A tree T equals its line-core if and only if T is a block-cutpoint tree.

$$(\text{Harary and Plummer } [HP19])$$

10.17   For any graph G, the following are equivalent:

1. G has a line-core,
2. $\alpha_0(G) = \beta_1(G)$,
3. $\alpha_1(G) = \beta_0(G)$.   (Harary and Plummer $[HP19]$)

10.18   If G is a connected graph having a line-core $C_1(G)$, then

a) $C_1(G)$ is a spanning subgraph of G,
b) $C_1(C_1(G)) = C_1(G)$,
c) the components of $C_1(G)$ are bipartite subgraphs of G which are not reducible.

$$(\text{Harary and Plummer } [HP19])$$

10.19   If G is a graph with line-core $C_1(G)$ and B is a bipartite subgraph of G properly containing $C_1(G)$, then B is reducible.   (Harary and Plummer $[HP\ 19]$)

10.20   The *point-core* $C_0(G)$ is the subgraph of G induced by the union of all independent sets S of $\alpha_1(G)$ points.   A graph G has a point-core if and only if it has a line-core.

$$(\text{Harary and Plummer } [HP\ 18])$$

10.21   If $G = C_0(G)$, then G has a 1-factor.   (Harary and Plummer $[HP\ 18]$)

10.22   If G is regular of degree n, then there is a partition of V into at most $1 + [n/2]$ subsets such that each point is adjacent to at most one other point in the same subset.

$$(\text{Gerencsér } [G6])$$

CHAPTER 11

PLANARITY

Return with me a while to the plains of Flatland,
and I will shew you that which you have often
reasoned and thought about . . .

EDWIN A. ABBOTT, *Flatland*

Topological graph theory was first discovered in 1736 by Euler ($V - E + F = 2$) and then was dormant for 191 years. The subject was revived when Kuratowski found a criterion for a graph to be planar. Another pioneer in topological graph theory was Whitney, who developed some important properties of the embedding of graphs in the plane.

All the known criteria for planarity are presented. These include the theorems of Kuratowski and Wagner, which characterize planar graphs in terms of forbidden subgraphs, Whitney's result in terms of the existence of a combinatorial dual, and MacLane's description of the existence of a prescribed cycle basis.

Several topological invariants of a graph are introduced. The genus of a graph has been determined for the complete graphs and bipartite graphs, the thickness for "most" of them, and the crossing number for only a few.

PLANE AND PLANAR GRAPHS

A graph is said to be *embedded* in a surface S when it is drawn on S so that no two edges intersect. As noted in Chapter 1, we shall use "points and lines" for abstract graphs, "vertices and edges" for geometric graphs (embedded in some surface). A graph is *planar* if it can be embedded in the plane; a

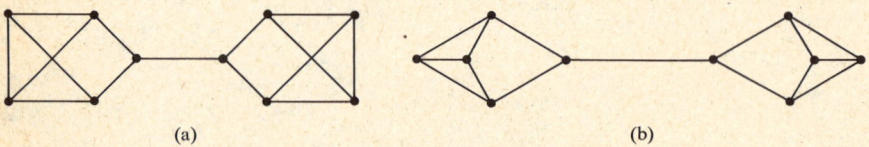

(a)   (b)

Fig. 11.1. A planar graph and an embedding.

plane graph has already been embedded in the plane. For example, the cubic graph of Fig. 11.1(a) is planar since it is isomorphic to the plane graph in Fig. 11.1(b).

We will refer to the regions defined by a plane graph as its *faces*, the unbounded region being called the *exterior face*. When the boundary of a face of a plane graph is a cycle, we will sometimes refer to the cycle as a face. The plane graph of Fig. 11.2 has three faces, f_1, f_2 and the exterior face f_3. Of these, only f_2 is bounded by a cycle.

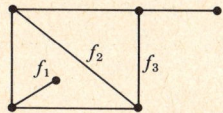

Fig. 11.2. A plane graph.

The subject of planar graphs was discovered by Euler in his investigation of polyhedra. With every polyhedron there is associated a graph consisting only of its vertices and edges, called its *1-skeleton*. For example, the graph Q_3 is the 1-skeleton of the cube and $K_{2,2,2}$ that of the octahedron. The Euler formula for polyhedra is one of the classical results of mathematics.

Theorem 11.1 (Euler Polyhedron Formula). For any spherical polyhedron with V vertices, E edges, and F faces,

$$V - E + F = 2. \tag{11.1}$$

For the 3-cube we have $V = 8$, $E = 12$, and $F = 6$ so that (11.1) holds; for a tetrahedron, $V = F = 4$ and $E = 6$. Before proving this equation, we will recast it in graph theoretic terms. A *plane map* is a connected plane graph together with all its faces. One can restate (11.1) for a plane map in terms of the numbers p of vertices, q of edges, and r of faces,

$$p - q + r = 2. \tag{11.1'}$$

It is easy to prove this theorem by induction. However, this equation has already been proved in Chapter 4 where it was established that the cycle rank m of a connected graph G is given by

$$m = q - p + 1.$$

Since it is easily seen that if (11.1') holds for the blocks of G separately, then (11.1') holds for G also, we assume from the outset that G is 2-connected. Thus every face of a plane embedding of G is a cycle.

We have just noted that $p = V$ and $q = E$ for a plane map. It only remains to link m with F. We now show that the interior faces of a plane graph G constitute a cycle basis for G, so that they are m in number. This holds because the edges of every cycle Z of G can be regarded as the symmetric difference of the faces of G contained in Z. Since the exterior face is thus the sum (mod 2) of all the interior faces (regarded as edge sets), we see that $m = F - 1$. Hence $m = q - p + 1$ becomes $F - 1 = E - V + 1$.

Euler's equation has many consequences.

Corollary 11.1(a) If G is a (p, q) plane map in which every face is an n-cycle, then

$$q = n(p - 2)/(n - 2). \tag{11.2}$$

Proof. Since every face of G is an n-cycle, each line of G is on two faces and each face has n edges. Thus $nr = 2q$, which when substituted into (11.1') gives the result.

A *maximal planar graph* is one to which no line can be added without losing planarity. Substituting $n = 3$ and 4 into (11.2) gives us the next result.

Corollary 11.1(b) If G is a (p, q) maximal plane graph, then every face is a triangle and $q = 3p - 6$. If G is a plane graph in which every face is a 4-cycle, then $q = 2p - 4$.

Because the maximum number of edges in a plane graph occurs when each face is a triangle, we obtain a necessary condition for planarity of a graph in terms of the number of lines.

Corollary 11.1(c) If G is any planar (p, q) graph with $p \geq 3$, then $q \leq 3p - 6$. Furthermore, if G has no triangles, then $q \leq 2p - 4$.

Corollary 11.1(d) The graphs K_5 and $K_{3,3}$ are nonplanar.

Proof. The $(5, 10)$ graph K_5 is nonplanar because $q = 10 > 9 = 3p - 6$; for $K_{3,3}$, $q = 9$ and $2p - 4 = 8$.

As we will soon see, the graphs K_5 and $K_{3,3}$ play a prominent role in characterizing planarity. The above corollaries are extremely useful in investigating planar graphs, especially maximal planar graphs.

Corollary 11.1(e) Every planar graph G with $p \geq 4$ has at least four points of degree not exceeding 5.

Clearly, a graph is planar if and only if each of its components is planar. Whitney [W12] showed that in studying planarity, it is sufficient to consider 2-connected graphs.

Theorem 11.2 A graph is planar if and only if each of its blocks is planar.

It is intuitively obvious that any planar graph can be embedded in the sphere, and conversely. This fact enables us to embed a planar graph in the plane in many different ways.

Theorem 11.3 Every 2-connected plane graph can be embedded in the plane so that any specified face is the exterior.

Proof. Let *f* be a nonexterior face of a plane block *G*. Embed *G* on a sphere and call some point interior to *f* the "North Pole." Consider a plane tangent to the sphere at the South Pole and project* *G* onto that plane from the North Pole. The result is a plane graph isomorphic to *G* in which *f* is the exterior face.

Corollary 11.3(a) Every planar graph can be embedded in the plane so that a prescribed line is an edge of the exterior region.

Whitney also proved that every maximal planar graph is a block, and more.

Theorem 11.4 Every maximal planar graph with $p \geq 4$ points is 3-connected.

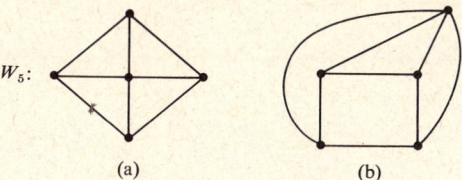

(a)   (b)

Fig. 11.3. Plane wheels.

There are five ways of embedding the 3-connected wheel W_5 in the plane: one looks like Fig. 11.3(a), and the other four look like Fig. 11.3(b). However, there is only one way of embedding W_5 on a sphere, an observation which holds for all 3-connected graphs (Whitney [W11]).

Theorem 11.5 Every 3-connected planar graph is uniquely embeddable on the sphere.

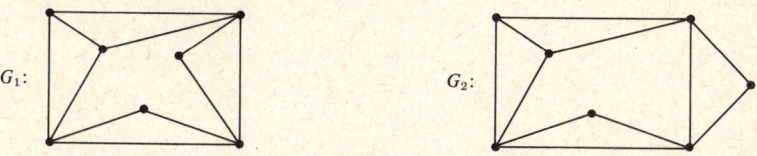

Fig. 11.4. Two plane embeddings of a 2-connected graph.

To show the necessity of 3-connectedness, consider the isomorphic graphs G_1 and G_2 of connectivity 2 shown in Fig. 11.4. The graph G_1 is embedded on the sphere so that none of its regions are bounded by five edges while G_2 has two regions bounded by five edges.

---

* This is usually called stereographic projection.

A polyhedron is *convex* if the straight line segment joining any two of its points lies entirely within it. The next theorem is due to Steinitz and Rademacher [SR2].

Theorem 11.6 A graph is the 1-skeleton of a convex 3-dimensional polyhedron if and only if it is planar and 3-connected.

One of the most fascinating areas of study in the theory of planar graphs is the interplay between considering a graph as a combinatorial object and as a geometric figure. Very often the question arises of placing geometric constraints on a graph. For example, Wagner [W1], Fáry [F1], and Stein [S15] independently showed that every planar graph can be embedded in the plane with straight edges.

Theorem 11.7 Every planar graph is isomorphic with a plane graph in which all edges are straight line segments.

OUTERPLANAR GRAPHS

A planar graph is *outerplanar* if it can be embedded in the plane so that all its vertices lie on the same face; we usually choose this face to be the exterior. Figure 11.5 shows an outerplanar graph (a) and two outerplane embeddings (b) and (c). In (c) all vertices lie on the exterior face.

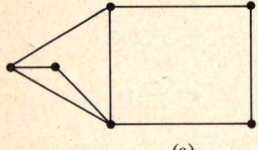
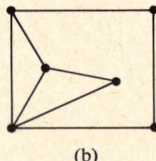
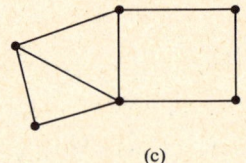

(a)   (b)   (c)

Fig. 11.5. An outerplanar graph and two outerplane embeddings.

In this section we develop theorems for outerplanar graphs parallel with those for planar graphs. The analogue of Theorem 11.2 is immediate.

Theorem 11.8 A graph G is outerplanar if and only if each of its blocks is outerplanar.

An outerplanar graph G is *maximal outerplanar* if no line can be added without losing outerplanarity. Clearly, every maximal outerplane graph is a triangulation of a polygon, while every maximal plane graph is a triangulation of the sphere. The three maximal outerplane graphs with 6 vertices are shown in Fig. 11.6.

Theorem 11.9 Let G be a maximal outerplane graph with $p \geq 3$ vertices all lying on the exterior face. Then G has $p - 2$ interior faces.

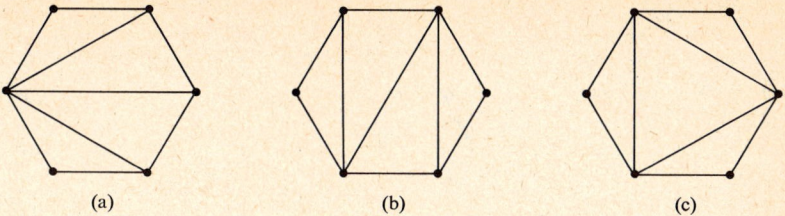

Fig. 11.6. Three maximal outerplanar graphs.

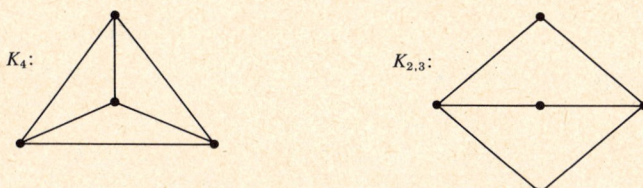

Fig. 11.7. The forbidden graphs for outerplanarity.

Proof. Obviously the result holds for $p = 3$. Suppose it is true for $p = n$ and let G have $p = n + 1$ vertices and m interior faces. Clearly G must have a vertex v of degree 2 on its exterior face. In forming $G - v$ we reduce the number of interior faces by 1 so that $m - 1 = n - 2$. Thus $m = n - 1 = p - 2$, the number of interior faces of G.

This theorem has several consequences.

Corollary 11.9(a) Every maximal outerplanar graph G with p points has

a) $2p - 3$ lines,
b) at least three points of degree not exceeding 3,
c) at least two points of degree 2,
d) $\kappa(G) = 2$.

All plane embeddings of K_4 and $K_{2,3}$ are of the forms shown in Fig. 11.7, in which each has a vertex inside the exterior cycle. Therefore, neither of these graphs is outerplanar. We now observe that these are the two basic non-outerplanar graphs, following [CH3].

Two graphs are *homeomorphic* if both can be obtained from the same graph by a sequence of subdivisions of lines. For example, any two cycles are homeomorphic, and Fig. 11.8 shows a homeomorph of K_4.

Theorem 11.10 A graph is outerplanar if and only if it has no subgraph homeomorphic to K_4 or $K_{2,3}$ except $K_4 - x$.

It is often important to investigate the complement of a graph with a given property. For planar graphs, the following theorem due to Battle,

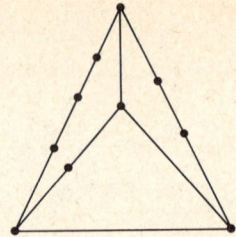

Fig. 11.8. A homeomorph of K_4.

Harary, and Kodama [BHK1] and proved less clumsily by Tutte [T16], provides a sufficient condition for the complement of a planar graph to be planar.

Theorem 11.11 Every planar graph with at least nine points has a nonplanar complement, and nine is the smallest such number.

This result was proved by exhaustion; no elegant or even reasonable proof is known.

The analogous observation for outerplanar graphs was made in [G5].

Theorem 11.12 Every outerplanar graph with at least seven points has a nonouterplanar complement, and seven is the smallest such number.

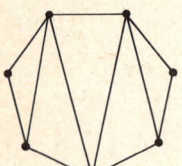 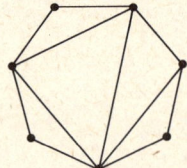 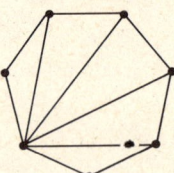 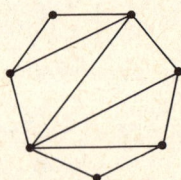

Fig. 11.9. The four maximal outerplanar graphs with seven points.

Proof. To prove the first part, it is sufficient to verify that the complement of every maximal outerplanar graph with seven points is not outerplanar. This holds because there are exactly four maximal outerplanar graphs with $p = 7$ (Fig. 11.9) and the complement of each is readily seen to be nonouterplanar. The minimality follows from the fact that the (maximal) outerplanar graph of Fig. 11.6(b) with six points has an outerplanar complement.

KURATOWSKI'S THEOREM

Until Kuratowski's paper appeared [K14], it was a tantalizing unsolved problem to characterize planar graphs. The following proof of his theorem is based on that by Dirac and Schuster [DS1].

Theorem 11.13 A graph is planar if and only if it has no subgraph homeomorphic to K_5 or $K_{3,3}$.

Proof. Since K_5 and $K_{3,3}$ are nonplanar by Corollary 11.1(d), it follows that if a graph contains a subgraph homeomorphic to either of these, it is also nonplanar.

The proof of the converse is a bit more involved. Assume it is false. Then there is a nonplanar graph with no subgraph homeomorphic to either K_5 or $K_{3,3}$. Let G be any such graph having the minimum number of lines. Then G must be a block with $\delta(G) \geq 3$. Let $x_0 = u_0 v_0$ be an arbitrary line of G. The graph $F = G - x_0$ is necessarily planar.

We will find it convenient to use two lemmas in the development of the proof.

Lemma 11.13(a) There is a cycle in F containing u_0 and v_0.

Proof of Lemma. Assume that there is no cycle in F containing u_0 and v_0. Then u_0 and v_0 lie in different blocks of F by Theorem 3.3. Hence, there exists a cutpoint w of F lying on every u_0–v_0 path. We form the graph F_0 by adding to F the lines wu_0 and wv_0 if they are not already present in F. In the graph F_0, u_0 and v_0 still lie in different blocks, say B_1 and B_2, which necessarily have the point w in common. Certainly, each of B_1 and B_2 has fewer lines than G, so either B_1 is planar or it contains a subgraph homeomorphic to K_5 or $K_{3,3}$. If, however, the insertion of wu_0 produces a subgraph H of B_1 homeomorphic to K_5 or $K_{3,3}$, then the subgraph of G obtained by replacing wu_0 by a path from u_0 to w which begins with x_0 is necessarily homeomorphic to H and so to K_5 or $K_{3,3}$, but this is a contradiction. Hence, B_1 and similarly B_2 is planar. According to Corollary 11.3(a), both B_1 and B_2 can be drawn in the plane so that the lines wu_0 and wv_0 bound the exterior region. Hence it is possible to embed the graph F_0 in the plane with both wu_0 and wv_0 on the exterior region. Inserting x_0 cannot then destroy the planarity of F_0. Since G is a subgraph of $F_0 + x_0$, G is planar; this contradiction shows that there is a cycle in F containing u_0 and v_0.

Let F be embedded in the plane in such a way that a cycle Z containing u_0 and v_0 has a maximum number of regions interior to it. Orient the edges of Z in a cyclic fashion, and let $Z[u, v]$ denote the oriented path from u to v along Z. If v does not immediately follow u on Z, we also write $Z(u, v)$ to indicate the subpath of $Z[u, v]$ obtained by removing u and v.

By the exterior of cycle Z, we mean the subgraph of F induced by the vertices lying outside Z, and the components of this subgraph are called the exterior components of Z. By an outer piece of Z, we mean a connected subgraph of F induced by all edges incident with at least one vertex in some exterior component or by an edge (if any) exterior to Z meeting two vertices of Z. In a like manner, we define the interior of cycle Z, interior component, and inner piece.

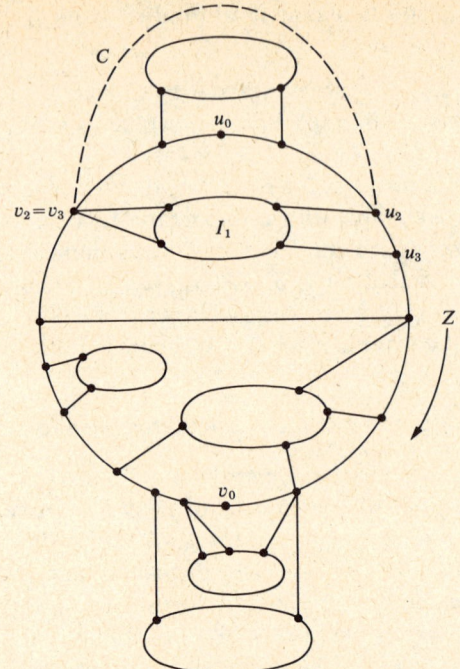

Fig. 11.10. Separating cycle Z illustrating lemma.

An outer or inner piece is called u–v separating if it meets both $Z(u, v)$ and $Z(v, u)$. Clearly, an outer or inner piece cannot be u–v separating if u and v are adjacent on Z.

Since F is connected, each outer piece must meet Z, and because F has no cutvertices, each outer piece must have at least two vertices in common with Z. No outer piece can meet $Z(u_0, v_0)$ or $Z(v_0, u_0)$ in more than one vertex, for otherwise there would exist a cycle containing u_0 and v_0 with more interior regions than Z. For the same reason, no outer piece can meet u_0 or v_0. Hence, every outer piece meets Z in exactly two vertices and is u_0–v_0 separating. Furthermore, since x_0 cannot be added to F in planar fashion, there is at least one u_0–v_0 separating inner piece.

Lemma 11.13(b) There exists a u_0–v_0 separating outer piece meeting $Z(u_0, v_0)$, say at u_1, and $Z(v_0, u_0)$, say at v_1, such that there is an inner piece which is both u_0–v_0 separating and u_1–v_1 separating.

Proof of Lemma. Suppose, to the contrary, that the lemma does not hold. It will be helpful in understanding this proof to refer to Fig. 11.10.

We order the u_0–v_0 separating inner pieces for the purpose of relocating them in the plane. Consider any u_0–v_0 separating inner piece I_1 which is

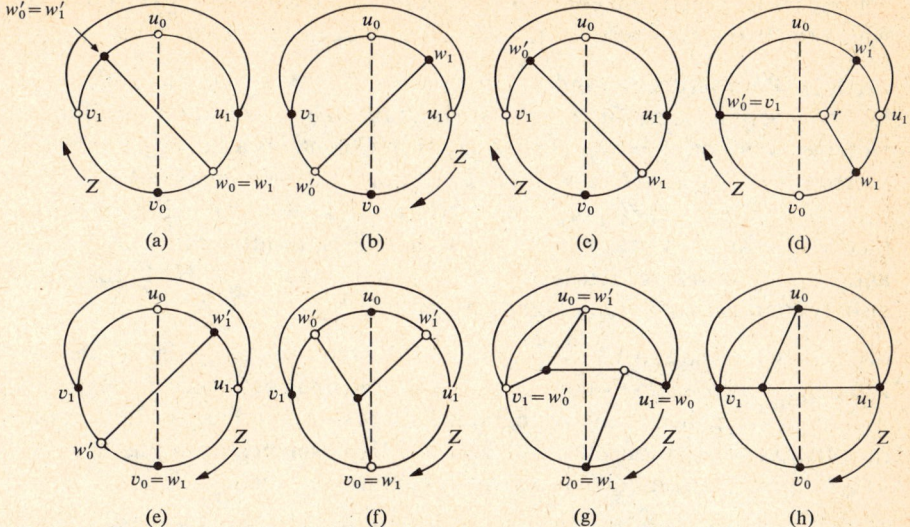

Fig. 11.11. The possibilities for nonplanar subgraphs.

nearest to u_0 in the sense of encountering points of this inner piece on moving along Z from u_0. Continuing out from u_0, we can index the u_0–v_0 separating inner pieces I_2, I_3, and so on.

Let u_2 and u_3 be the first and last points of I_1 meeting $Z(u_0, v_0)$ and v_2 and v_3 be the first and last vertices of I_1 meeting $Z(v_0, u_0)$. Every outer piece necessarily has both its common vertices with Z on either $Z[v_3, u_2]$ or $Z[u_3, v_2]$, for otherwise there would exist an outer piece meeting $Z(u_0, v_0)$ at u_1 and $Z(v_0, u_0)$ at v_1 and an inner piece which is both u_0–v_0 separating and u_1–v_1 separating, contrary to the supposition that the lemma is false. Therefore, a curve C joining v_3 and u_2 can be drawn in the exterior region so that it meets no edge of F. (See Fig. 11.10.) Thus, I_1 can be transferred outside of C in a planar manner. Similarly, the remaining u_0–v_0 separating inner pieces can be transferred outside of Z, in order, so that the resulting graph is plane. However, the edge x_0 can then be added without destroying the planarity of F, but this is a contradiction, completing the lemma.

Proof of Theorem. Let H be the inner piece guaranteed by Lemma 11.13(b) which is both u_0–v_0 separating and u_1–v_1 separating. In addition, let w_0, w_0', w_1, and w_1' be vertices at which H meets $Z(u_0, v_0), Z(v_0, u_0), Z(u_1, v_1)$, and $Z(v_1, u_1)$, respectively. There are now four cases to consider, depending on the relative position on Z of these four vertices.

CASE 1. One of the vertices w_1 and w_1' is on $Z(u_0, v_0)$ and the other is on $Z(v_0, u_0)$. We can then take, say, $w_0 = w_1$ and $w_0' = w_1'$, in which case G

contains a subgraph homeomorphic to $K_{3,3}$, as indicated in Fig. 11.11(a), in which the two sets of vertices are indicated by open and closed dots.

CASE 2.   Both vertices w_1 and w_1' are on either $Z(u_0, v_0)$ or $Z(v_0, u_0)$. Without loss of generality we assume the first situation. There are two possibilities: either $v_1 \neq w_0'$ or $v_1 = w_0'$. If $v_1 \neq w_0'$, then G contains a subgraph homeomorphic to $K_{3,3}$, as shown in Fig. 11.11(b) or (c), depending on whether w_0' lies on $Z(u_1, v_1)$ or $Z(v_1, u_1)$, respectively.   If $v_1 = w_0'$ (see Fig. 11.11d), then H contains a vertex r from which there exist disjoint paths to w_1, w_1', and v_1, all of whose vertices (except w_1, w_1', and v_1) belong to H.   In this case also, G contains a subgraph homeomorphic to $K_{3,3}$.

CASE 3. $w_1 = v_0$ and $w_1' \neq u_0$. Without loss of generality, let w_1' be on $Z(u_0, v_0)$. Once again G contains a subgraph homeomorphic to $K_{3,3}$.   If w_0' is on (v_0, v_1), then G has a subgraph $K_{3,3}$ as shown in Fig. 11.11(e). If, on the other hand, w_0' is on $Z(v_1, u_0)$, there is a $K_{3,3}$ as indicated in Fig. 11.11(f). This figure is easily modified to show G contains $K_{3,3}$ if $w_0' = v_1$.

CASE 4. $w_1 = v_0$ and $w_1' = u_0$. Here we assume $w_0 = u_1$ and $w_0' = v_1$, for otherwise we are in a situation covered by one of the first 3 cases.   We distinguish between two subcases. Let P_0 be a shortest path in H from u_0 to v_0, and let P_1 be such a path from u_1 to v_1. The paths P_0 and P_1 must intersect. If P_0 and P_1 have more than one vertex in common, then G contains a subgraph homeomorphic to $K_{3,3}$, as shown in Fig. 11.11(g); otherwise, G contains a subgraph homeomorphic to K_5 as in Fig. 11.11(h).

Since these are all the possible cases, the theorem has been proved.

In his paper "How to draw a graph," Tutte [T17] gives an algorithm for drawing in the plane as much of a given graph as possible and shows that whenever this process stops short of the entire graph, it must contain a subgraph homeomorphic to K_5 or $K_{3,3}$. Thus his algorithm furnishes an independent proof of Theorem 11.13.

An *elementary contraction* of a graph G is obtained by identifying two adjacent points u and v, that is, by the removal of u and v and the addition

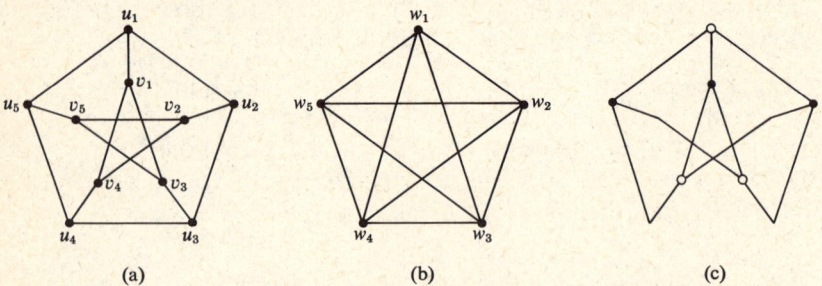

Fig. 11.12.   Nonplanarity of the Petersen graph.

of a new point w adjacent to those points to which u or v was adjacent. A graph G is *contractible* to a graph H if H can be obtained from G by a sequence of elementary contractions. For example, as indicated in Fig. 11.12(a) and (b), the Petersen graph is contractible to K_5 by contracting each of the five lines u_iv_i joining the pentagon with the pentagram to a new point w_i. A dual form of Kuratowski's theorem (in the sense of duality in matroid theory) was found independently by Wagner [W2] and Harary and Tutte [HT3].

Theorem 11.14 A graph is planar if and only if it does not have a subgraph contractible to K_5 or $K_{3,3}$.

We have just seen that the Petersen graph is contractible to K_5. Since every point has degree 3, it clearly does not have a subgraph homeomorphic to K_5; Fig. 11.12(c) shows one homeomorphic to $K_{3,3}$.

OTHER CHARACTERIZATIONS OF PLANAR GRAPHS

Several other criteria for planarity have been discovered since the original work of Kuratowski. We have already noted the "dual form" in terms of contraction in Theorem 11.14. Tutte's algorithm for drawing a graph in the plane may also be regarded as a characterization.

Whitney [W12, W14] expressed planarity in terms of the existence of dual graphs. Given a plane graph G, its *geometric dual* G^* is constructed as follows: place a vertex in each region of G (including the exterior region) and, if two regions have an edge x in common, join the corresponding vertices by an edge x^* crossing only x. The result is always a plane pseudograph, as indicated in Fig. 11.13 where G has solid edges and its dual G^* dashed edges. Clearly G^* has a loop if and only if G has a bridge, and G^* has multiple edges if and only if two regions of G have at least two edges in common. Thus, a 2-connected plane graph always has a graph or multigraph as its dual, while the dual of a 3-connected graph is always a graph. Other examples of geometric duals are given by the Platonic graphs: the tetrahedron is self-dual, whereas the cube and octahedron are duals, as are the dodecahedron and the icosahedron.

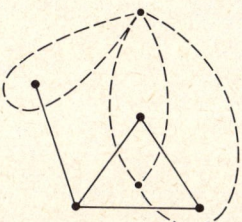

Fig. 11.13. A plane graph and its geometric dual.

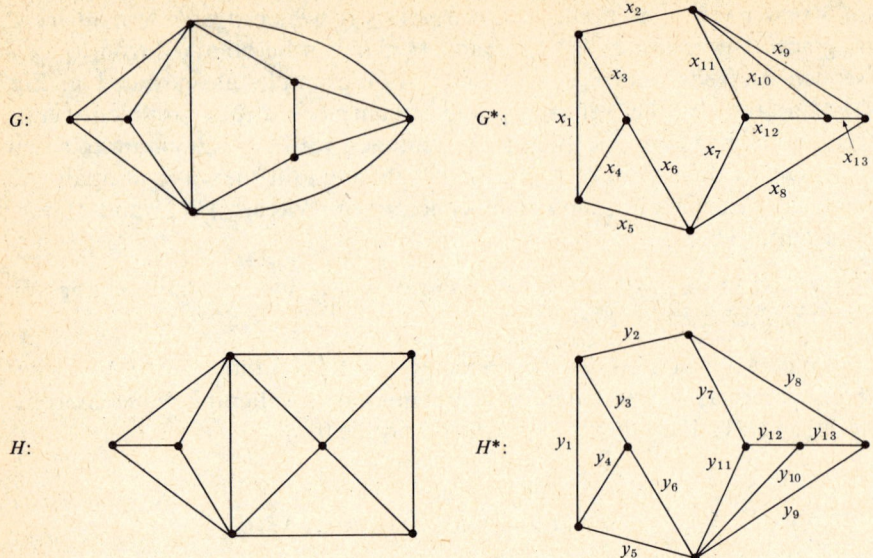

Fig. 11.14. Different geometric duals of the same abstract graph.

As defined, the geometric dual of a connected plane graph G is also plane, and it follows that the dual of the dual of G is the original graph G. However, an abstract graph with more than one embedding on the sphere can give rise to more than one dual graph. This is illustrated in Fig. 11.14 in which graphs G and H are abstractly isomorphic, but as embedded they have different duals G^* and H^*. However, since a triply connected graph has only one spherical embedding, as noted in Theorem 11.5, it must have a unique geometric dual.

Whitney gave a combinatorial definition of dual, which is an abstract formulation of the geometric dual. To state this, we recall from Chapter 4 that for a graph G with k components, the cycle rank is given by $m(G) = q - p + k$ and the cocycle rank by $m^*(G) = p - k$.

The relative complement $G - H$ of a subgraph H of G is defined to be that subgraph obtained by deleting the lines of H. A graph G^* is a *combinatorial dual* of graph G if there is a one-to-one correspondence between their sets of lines such that for any choice Y and Y^* of corresponding subsets of lines,

$$m^*(G - Y) = m^*(G) - m(\langle Y^* \rangle), \tag{11.3}$$

where $\langle Y^* \rangle$ is the subgraph of G^* with line set Y^*. This definition is illustrated by Fig. 11.15 where the correspondence is $x_i \leftrightarrow y_i$. Here $Y = \{x_2, x_3, x_4, x_6\}$, so that $m^*(G - Y) = 4$, $m^*(G) = 5$, and $m(\langle Y^* \rangle) = 1$, so

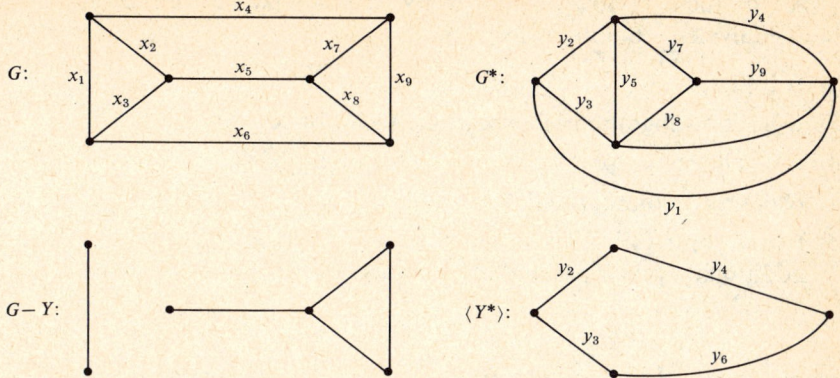

Fig. 11.15. Combinatorial duals.

the defining equation is satisfied. It is of course very difficult to check whether two graphs are duals using (11.3) since it involves verifying this equation for every set Y of lines in G.

As with geometric duals, combinatorial duals of planar graphs are not necessarily unique. However, if two graphs are combinatorial duals of isomorphic graphs, there is a one-to-one correspondence between their sets of lines which preserves cycles as sets of lines (that is, their cycle matroids are isomorphic). The correspondence $x_i \leftrightarrow y_i$ of G^* and H^* in Fig. 11.14 illustrates this.

Whitney proved that combinatorial duals are equivalent to geometric duals, giving another criterion for planarity.

Theorem 11.15 A graph is planar if and only if it has a combinatorial dual.

Another criterion for planarity due to MacLane [M1] is expressed in terms of cyclic structure.

Theorem 11.16 A graph G is planar if and only if every block of G with at least three points has a cycle basis $Z_1, Z_2, \cdots, Z_m$ and one additional cycle Z_0 such that every line occurs in exactly two of these $m + 1$ cycles.

We only indicate the necessity, which is much easier. As mentioned in the proof of Theorem 11.1, all the interior faces of a 2-connected plane graph G constitute a cycle basis $Z_1, Z_2, \cdots, Z_m$, where m is the cycle rank of G. Let Z_0 be the exterior cycle of G. Then obviously each edge of G lies on exactly two of the $m + 1$ cycles Z_i.

To prove the sufficiency, it is necessary to construct a plane embedding of a given graph G with the stipulated properties.

All of these criteria for planarity are summarized in the following list of equivalent conditions for a graph G.

(1) G is planar.

(2) G has no subgraph homeomorphic to K_5 or $K_{3,3}$.

(3) G has no subgraph contractible to K_5 or $K_{3,3}$.

(4) G has a combinatorial dual.

(5) Every nontrivial block of G has a cycle basis $Z_1, Z_2, \cdots, Z_m$ and one additional cycle Z_0 such that every line x occurs in exactly two of these $m + 1$ cycles.

GENUS, THICKNESS, COARSENESS, CROSSING NUMBER

In this section four topological invariants of a graph G are considered. These are genus: the number of handles needed on a sphere in order to embed G, thickness: the number of planar graphs required to form G, coarseness: the maximum number of line-disjoint nonplanar subgraphs in G, and crossing number: the number of crossings there must be when G is drawn in the plane. We will concentrate on three classes of graphs—complete graphs, complete bigraphs, and cubes—and indicate the values of these invariants for them as far as they are known.

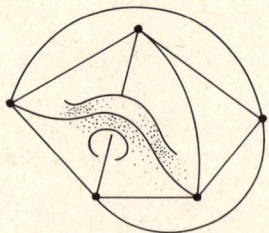

Fig. 11.16. Embedding a graph on an orientable surface.

As observed by König, every graph is embeddible on some orientable surface. This can easily be seen by drawing an arbitrary graph G in the plane, possibly with edges that cross each other, and then attaching a handle to the plane at each crossing and allowing one edge to go over the handle and the other under it. For example, Fig. 11.16 shows an embedding of K_5 in a plane to which one handle has been attached. Of course, this method often uses more handles than are actually required. In fact, König also showed that any embedding of a graph on an orientable surface with a minimum number of handles has all its faces simply connected.

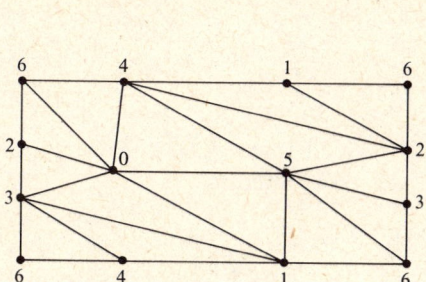

Fig. 11.17. An embedding of K_7 on the torus.

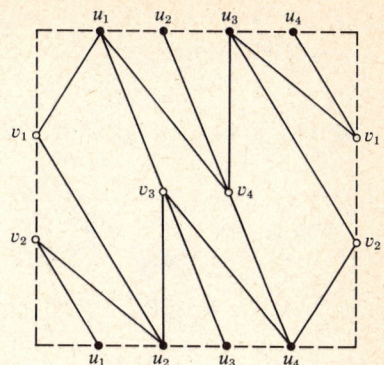

Fig. 11.18. A toroidal embedding of $K_{4,4}$.

We have already noted that planar graphs can be embedded on a sphere. A *toroidal graph* can be embedded on a torus. Both K_5 and $K_{3,3}$ are toroidal; in fact Figs. 11.17 and 11.18 show embeddings of K_7 and $K_{4,4}$ on the torus, represented as the familiar rectangle in which both pairs of opposite sides are identified. No characterization of toroidal graphs analogous to Kuratowski's Theorem has been found. However, Vollmerhaus [V6] settled a conjecture of Erdös in the affirmative by proving that for the torus as well as any other orientable surface, there is a finite collection of forbidden subgraphs.

The *genus* $\gamma(G)$ of a graph G is the minimum number of handles which must be added to a sphere so that G can be embedded on the resulting surface. Of course, $\gamma(G) = 0$ if and only if G is planar, and homeomorphic graphs have the same genus.

The first theorem of this chapter presented the Euler characteristic equation, $V - E + F = 2$, for spherical polyhedra. More generally, the *genus of a polyhedron** is the number of handles needed on the sphere for a surface to contain the polyhedron. Theorem 11.1 has been generalized to polyhedra of arbitrary genus, in a result also due to Euler. A proof may be found in Courant and Robbins [CR1].

Theorem 11.17 For a polyhedron of genus γ with V vertices, E edges and F faces,

$$V - E + F = 2 - 2\gamma. \qquad (11.4)$$

This equation is particularly useful in proving the easy half of the results to follow on the genus and thickness of particular graphs. Its corollaries, which offer no difficulty, are often more convenient for this purpose.

---

* For a combinatorial treatment of the theory of polyhedra, see Grünbaum [G10].

Corollary 11.17(a) If G is a connected graph of genus γ in which every face is a triangle, then

$$q = 3(p - 2 + 2\gamma); \tag{11.5}$$

when every face is a quadrilateral,

$$q = 2(p - 2 + 2\gamma). \tag{11.6}$$

As mentioned in [BH2], it is easily verified from these two equations that the genus of a graph has the following lower bounds.

Corollary 11.17(b) If G is a connected graph of genus γ, then

$$\gamma \geq \tfrac{1}{6}q - \tfrac{1}{2}(p - 2); \tag{11.7}$$

if G has no triangles, then

$$\gamma \geq \tfrac{1}{4}q - \tfrac{1}{2}(p - 2). \tag{11.8}$$

The determination of the genus of the complete graphs has been a long, interesting, difficult, successful struggle. In its dual form, it was known as the Heawood Conjecture and stood unproved from 1890 to 1967. We return to this aspect of the problem in the next chapter. There have been many contributors to this result and the *coup de grace*, settling the conjecture in full, was administered by Ringel and Youngs [RY1].

Theorem 11.18 For $p \geq 3$, the genus of the complete graph is

$$\gamma(K_p) = \left\{ \frac{(p - 3)(p - 4)}{12} \right\}. \tag{11.9}$$

The proof of the easier half of equation (11.9) is due to Heawood [H38]. It amounts to substituting $q(K_p)$ into inequality (11.7) to obtain

$$\gamma(K_p) \geq \frac{1}{6}\binom{p}{2} - \frac{1}{2}(p - 2) = \frac{(p - 3)(p - 4)}{12}.$$

Then since the genus of every graph is an integer,

$$\gamma(K_p) \geq \left\{ \frac{(p - 3)(p - 4)}{12} \right\}.$$

The proof that this expression is also an upper bound for $\gamma(K_p)$ can only be accomplished by displaying an embedding of K_p into an orientable surface of the indicated genus. When Heawood originally stated the conjecture in 1890, he proved that $\gamma(K_7) = 1$, as verified by the embedding shown in Fig. 11.17, which triangulates the torus.

Heffter proved (11.9) in 1891 for $p = 8$ through 12. Not until 1952 did Ringel prove it for $p = 13$. At that stage, it was realized that because of its form, it was natural to try to settle the question for one residue class of p

modulo 12 at a time. Writing $p = 12s + r$, Ringel (see [R10]) proved (11.9) in 1954 for all complete graphs K_p with $r = 5$. During 1961–65, Ringel extended the result to $r = 7, 10$, and 3, and concurrently Youngs [Y1] with his colleagues Gustin, Terry, and Welch settled the cases $r = 4, 0, 1, 9, 6$. In 1967–68, Ringel and Youngs [RY1, 2] worked together to achieve appropriate embeddings of K_p for $r = 2, 8$, and 11. The isolated cases $p = 18$, 20, and 23 remained unproved by these methods. The proof was completed by the Professor of French Literature at the University of Montpellier, named Jean Mayer, when he embedded K_p for these three values of p, see [M6].

For complete bigraphs, the corresponding result is less involved, and was obtained by Ringel alone. Since inequality (11.8) applies to the graph $K_{m,n}$ we have

$$\gamma(K_{m,n}) \geq \frac{1}{4} mn - \frac{1}{2}(m + n - 2) = \frac{(m - 2)(n - 2)}{4}.$$

The other inequality is demonstrated [R12] by displaying a suitable embedding of $K_{m,n}$.

Theorem 11.19 The genus of the complete bigraph is

$$\gamma(K_{m,n}) = \left\{ \frac{(m - 2)(n - 2)}{4} \right\}. \tag{11.10}$$

The genus of the cube was derived by Ringel [R13] and Beineke and Harary [BH3]. For the graph Q_n, we have $p = 2^n$ and $q = n2^{n-1}$, so that by (11.8),

$$\gamma(Q_n) \geq 1 + (n - 4)2^{n-3},$$

proving the easier half of the next equation.

Theorem 11.20 The genus of the cube is

$$\gamma(Q_n) = 1 + (n - 4)2^{n-3}. \tag{11.11}$$

We now mention some more general considerations involving genus. It was shown in Battle, Harary, Kodama and Youngs [BHKY1] that the genus of a graph depends only on the genus of its blocks, as anticipated in Theorem 11.2.

Theorem 11.21 If a graph G has blocks $B_1, B_2, \cdots, B_n$, then

$$\gamma(G) = \sum_{i=1}^{n} \gamma(B_i). \tag{11.12}$$

This result was generalized slightly by Harary and Kodama [HK1]. Recall from Theorem 5.8 that two $(n + 1)$-components of a graph have at most n points in common.

Theorem 11.22 Let an n-connected graph G be the union of two $(n + 1)$-components B and C. Let $v_1, \cdots, v_n$ be the set of points of $B \cap C$. Call G_{ij} the graph obtained by adding line $v_i v_j$ to G. If $\gamma(G_{ij}) = \gamma(G) + 1$ whenever $1 \leq i < j \leq n$, then

$$\gamma(G) = \gamma(B) + \gamma(C) + n - 1. \tag{11.13}$$

We have already observed in Theorem 11.11 that every planar graph with 9 points has a nonplanar complement. Define the *thickness* $\theta(G)$ of a graph as the minimum number of planar subgraphs whose union is G. Then Theorem 11.11 can be stated in the form $\theta(K_9) > 2$. Actually the thickness of K_9 is 3 but K_9 is critical with respect to thickness since $\theta(K_9 - x) = 2$. Therefore $\theta(K_p) = 2$ for $p = 5$ to 8. Of course $\theta(G) = 1$ if and only if G is planar. Since a maximal planar graph has $q = 3p - 6$ lines, it follows that the thickness θ of any (p, q) graph has the bound,

$$\theta \geq \frac{q}{3p - 6}. \tag{11.14}$$

This observation is useful in making conjectures about thickness and proving the easier half.

The thickness of the complete graphs was investigated in [BH5] and Beineke [B6]. Applying (11.14) to K_p, we find

$$\theta(K_p) \geq \frac{p(p - 1)/2}{3(p - 2)}.$$

Applying some algebraic manipulations, we obtain

$$\theta(K_p) \geq \left\lceil \frac{p(p - 1)/2 + 3(p - 2) - 1}{3(p - 2)} \right\rceil = \left\lceil \frac{p + 7}{6} \right\rceil.$$

Theorem 11.23 Whenever $p \not\equiv 4 \pmod 6$, the thickness of the complete graph is

$$\theta(K_p) = \left\lceil \frac{p + 7}{6} \right\rceil \tag{11.15}$$

unless $p = 9$.

When $p \equiv 4 \pmod 6$, sometimes equation (11.15) holds and sometimes it doesn't. For $\theta(K_{10}) = 3 \neq \left\lceil \frac{17}{6} \right\rceil$, but Hobbs and Grossman [HG1] produced a decomposition of K_{22} into $4 = \left\lceil \frac{29}{6} \right\rceil$ planar subgraphs and Beineke [B6] showed that $\theta(K_{28}) = 5 = \left\lceil \frac{35}{6} \right\rceil$. Very recently, Jean Mayer (again!) obtained constructions showing that $\theta(K_{34}) = 6$ and $\theta(K_{40}) = 7$. The only value of $p \leq 45$ for which $\theta(K_p)$ is not yet known is $p = 16$. It is conjectured that $\theta(K_{16}) = 4$, but for all $p \equiv 4 \pmod 6$, and $p \geq 46$, that (11.15) holds.

The thickness of complete bigraphs was studied in [BHM1] and Beineke [B7].

Theorem 11.24 The thickness of the complete bigraph is

$$\theta(K_{m,n}) = \left\{\frac{mn}{2(m + n - 2)}\right\} \tag{11.16}$$

except possibly when $m < n$, mn is odd, and there exists an integer k such that $n = [2k(m - 2)/(m - 2k)]$.

Corollary 11.24(a) The thickness of $K_{n,n}$ is $[(n + 5)/4]$.

The corresponding problem for the cube was settled by Kleinert [K8].

Theorem 11.25 The thickness of the cube is

$$\theta(Q_n) = \left\{\frac{n + 1}{4}\right\}. \tag{11.17}$$

P. Erdös (verbal communication) made a fortuitous slip, while trying to describe the concept of thickness. By speaking of the maximum number of line-disjoint nonplanar subgraphs contained in the given graph G, he first defined the *coarseness* $\xi(G)$. Thus both thickness and coarseness involve constructions which factor a graph into spanning subgraphs (planar and nonplanar respectively) in the sense of Chapter 9. Formulas for the coarseness of a complete graph are not as neat as those for other topological invariants. The reason is that $K_{3,3}$ or a homeomorph thereof is a most convenient subgraph for coarseness constructions. This suggests the reason for the form of the next result due to Guy and Beineke [GB1]. Figure 11.19 shows four line-disjoint homeomorphs of $K_{3,3}$ contained in K_{10}.

Theorem 11.26 The coarseness of the complete graphs is given by

$$\xi(K_{3n}) = \begin{cases} \dbinom{n}{2} & (p = 3n \leq 15), \\ \dbinom{n}{2} + \left[\dfrac{n}{5}\right] & (p = 3n \geq 30), \end{cases}$$

$$\xi(K_{3n+1}) = \binom{n}{2} + 2\left[\frac{n}{3}\right] \qquad \begin{array}{l}(p = 3n + 1 \geq 19 \\ \text{and } p \neq 9r + 7), \end{array}$$

$$\xi(K_{3n+2}) = \binom{n}{2} + \left[\frac{14n + 1}{15}\right]. \tag{11.18}$$

All of the values of $\xi(K_p)$ are either known exactly from (11.18) or have the value given in Table 11.1 or 1 greater; see [GB1].

For the coarseness of the complete bigraph, the results of Beineke and Guy [BG1] are incomplete and involve many cases.

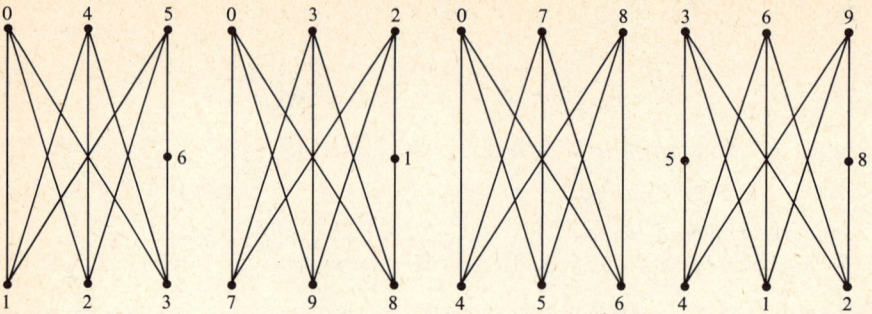

Fig. 11.19. Four nonplanar subgraphs of K_{10}.

Theorem 11.27 The coarseness of the complete bigraph $K_{m,n}$ satisfies

$$\zeta(K_{3r+d,3s+e}) = rs + \min\left(\left\lceil \frac{er}{3} \right\rceil, \left\lceil \frac{ds}{3} \right\rceil\right)$$

$$\text{for} \quad d = 0 \text{ or } 1 \quad \text{and} \quad e = 0 \text{ or } 1.$$

$$\zeta(K_{3r+2,3s}) = rs + \left\lceil \frac{s}{3} \right\rceil, \qquad \text{when } r \geq 1.$$

$$\zeta(K_{3r+2,3s+1}) \begin{cases} \leq rs + \min\left(\left\lceil \dfrac{r+s}{3} \right\rceil, \left\lceil \dfrac{2s}{3} \right\rceil, \left\lceil \dfrac{8r+16s+2}{39} \right\rceil\right) \\[3mm] \geq rs + \max\left(\left\lceil \dfrac{s+2}{3} \right\rceil, \min\left(\left\lceil \dfrac{r}{3} \right\rceil, \left\lceil \dfrac{2s}{3} \right\rceil\right)\right) \end{cases} \qquad (11.19)$$

$$\text{for} \quad r \geq 2, s \geq 7.$$

(These are equal when $r \geq 2s$.)

$$\zeta(K_{3r+2,3s+2}) \begin{cases} \leq rs + \min\left(\left\lceil \dfrac{r+2s}{3} \right\rceil, \left\lceil \dfrac{2r+s}{3} \right\rceil, \left\lceil \dfrac{16r+16s+4}{39} \right\rceil\right) \\[3mm] \geq rs + \left\lceil \dfrac{r}{3} \right\rceil + \left\lceil \dfrac{s}{3} \right\rceil + \left\lceil \dfrac{r}{9} \right\rceil \qquad \text{for} \quad 1 \leq r \leq s. \end{cases}$$

The *crossing number* $v(G)$ of a graph G is the minimum number of pairwise intersections of its edges when G is drawn in the plane. Obviously $v(G) = 0$ if and only if G is planar. The exact value of the crossing number has not yet been determined for any of the three families of graphs; only upper bounds are definitely established. The prevailing conjecture is that the bounds in (11.20) and (11.21) are exact. Several authors have deluded themselves into thinking they had proved equality. For details, see Guy [G12].

Table 11.1

CONJECTURED VALUES FOR $\xi(K_p)$

p	13	18	21	24	27	$9n + 7$
$\xi(K_p)$	7	15	21	28	36	$(9n^2 + 13n + 2)/2$

Theorem 11.28 The crossing number of the complete graph satisfies the inequality

$$v(K_p) \leq \frac{1}{4}\left[\frac{p}{2}\right]\left[\frac{p-1}{2}\right]\left[\frac{p-2}{2}\right]\left[\frac{p-3}{2}\right]. \tag{11.20}$$

Theorem 11.29 The crossing number of the complete bigraph satisfies the inequality

$$v(K_{m,n}) \leq \left[\frac{m}{2}\right]\left[\frac{m-1}{2}\right]\left[\frac{n}{2}\right]\left[\frac{n-1}{2}\right]. \tag{11.21}$$

T. Saaty showed that (11.20) is an equation for $p \leq 10$ while D. Kleitman proved equality in (11.21) for $m \leq 6$. These are the only known values of $v(K_p)$ and $v(K_{m,n})$. For the cubes, no one has even conjectured what is v.

EXERCISES

11.1 If a (p_1, q_1) graph and a (p_2, q_2) graph are homeomorphic, then

$$p_1 + q_2 = p_2 + q_1.$$

11.2 Every plane eulerian graph contains an eulerian trail that never crosses itself.

11.3 A 3-connected graph with $p \geq 6$ is planar if and only if no subgraph is homeomorphic to $K_{3,3}$. (D. W. Hall [H6])

*11.4 Every 4-connected planar graph is hamiltonian. (Tutte [T11½])

11.5 Every 5-connected planar graph has at least 12 points. Construct one.

11.6 There is no 6-connected planar graph.

*11.7 If G is a maximal plane graph in which every triangle bounds a region, then G is hamiltonian. (Whitney, *Ann. Math.*, **32** (1931), 378–390.)

11.8 Not every maximal planar graph is hamiltonian. (Whitney, same as above.)

11.9 If, in a drawing of G in the plane, every pair of nonadjacent edges cross an even number of times, then G is planar.

(R. L. Brooks, C. A. B. Smith, A. H. Stone, and W. T. Tutte)

11.10 Prove or disprove: every connected nonplanar graph has K_5 or $K_{3,3}$ as a contraction.

11.11 Prove or disprove: A graph is planar if and only if every subgraph with at most six points of degree at least 3 is homeomorphic to a subgraph of $K_2 + P_4$.

11.12 Prove or disprove: The cycle basis of a plane graph consisting of the interior faces always comes from a tree (cf. Chapter 4).

*11.13 Every triply connected planar graph has a spanning tree with maximum degree 3.

(Barnette [B3])

11.14 A plane graph is 2-connected if and only if its geometric dual is 2-connected.

11.15 All wheels are self-dual.

11.16 The square of a connected graph G is outerplanar if and only if G is K_3 or a path.

11.17 The following statements are equivalent:
   (1) The line graph $L(G)$ is outerplanar.
   (2) The maximum degree $\Delta(G) \le 3$ and every point of degree 3 is a cutpoint.
   (3) The total graph $T(G)$ is planar.

(Chartrand, Geller, and Hedetniemi [CGH2], Behzad [B4])

11.18 A graph G other than $\triangleleft\!\!\triangleright$ has a planar square if and only if $\Delta(G) \le 3$, every point of degree 3 is a cutpoint, and all blocks of G with more than 3 points are even cycles.

(Harary, Karp, and Tutte [HKT1])

11.19 A graph G has a planar line graph if and only if G is planar, $\Delta(G) \le 4$, and every point of degree 4 is a cutpoint. (Sedláček [S10])

11.20 Find the genus and crossing number of the Petersen graph.

11.21 Prove or disprove: A nonplanar graph G has $v = 1$ if and only if $G - x$ is planar for some line x.

11.22 The arboricity of every planar graph is at most 3. Construct a planar graph with arboricity 3.

11.23 Every graph is homeomorphic to a graph with arboricity 1 or 2, and hence of thickness 1 or 2.

11.24 The skewness of G is the minimum number of lines whose removal results in a planar graph. Find the skewness of
   a) K_p,   b) $K_{m,n}$,   c) Q_n. (A. Kotzig)

11.25 If G is outerplanar without triangles, then

$$q \le (3p - 4)/2.$$

11.26 If G is a graph such that for any two points, there are at most two point-disjoint paths of length greater than 1 joining them, then
   a) G is planar.
   b) $q \le 2p - 2$.
   c) If G is nonseparable and $p \ge 5$, then there is a unique hamiltonian cycle.

(Tang [T2])

11.27 Embed the cube Q_4 on the surface of a torus.

11.28   The genus γ of any graph G with girth g has the lower bound

$$\gamma \geq \frac{1}{2}\left[\left(1 - \frac{2}{g}\right)q - (p - 2)\right].$$

(Beineke and Harary [BH2])

*11.29 $\gamma(K_{n,n,n}) = \binom{n-1}{2}.$

(G. Ringel)

11.30   If G_1 and G_2 are homeomorphic, then $\xi(G_1) = \xi(G_2)$ and $v(G_1) = v(G_2)$.

11.31   The maximum number of line-disjoint $K_{3,3}$ subgraphs in $K_{m,n}$ is

$$\min\left(\left[\frac{m}{3}\left[\frac{n}{3}\right]\right], \left[\frac{n}{3}\left[\frac{m}{3}\right]\right]\right).$$

Thus for all n,

$$\xi(K_{n,n}) \geq \left[\frac{n}{3}\left[\frac{n}{3}\right]\right].$$

(Beineke and Guy [BG1])

CHAPTER 12

COLORABILITY

> Suppose there's a brown calf and a big brown dog, and an artist
> is making a picture of them . . . He has got to paint them so you can
> tell them apart the minute you look at them, hain't he? Of course.
> Well, then, do you want him to go and paint both of them brown?
> Certainly you don't. He paints one of them blue, and then you can't
> make no mistake. It's just the same with maps.
> That's why they make every state a different color . . .
>
> SAMUEL CLEMENS (MARK TWAIN)

The Four Color Conjecture (4CC) can truly be renamed the "Four Color Disease" for it exhibits so many properties of an infection. It is highly contagious. Some cases are benign and others malignant or chronic. There is no known vaccine, but men with a sufficiently strong constitution have achieved life-long immunity after a mild bout. It is recurrent and has been known to cause exquisite pain although there are no terminal cases on record. At least one case of the disease was transmitted from father to son, so it may be hereditary.

It is this problem which has stimulated results on colorability of graphs, which have led in turn to the investigation of several other areas of graph theory. After describing the coloring of a graph and its chromatic number, the stage is set for a proof of the Five Color Theorem and a discussion of the Four Color Conjecture. We then introduce uniquely colorable graphs, which can only be colored in one way, and critical graphs, which are minimal with respect to coloring. The intimate relationship between homomorphisms and colorings is investigated. The chapter concludes with a development of the properties of the chromatic polynomial.

THE CHROMATIC NUMBER

A *coloring* of a graph is an assignment of colors to its points so that no two adjacent points have the same color. The set of all points with any one color is independent and is called a *color class*. An *n-coloring* of a graph G uses n colors; it thereby partitions V into n color classes. The *chromatic*

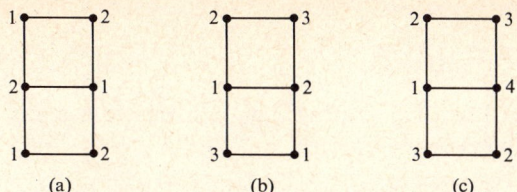

Fig. 12.1. Three colorings of a graph.

number $\chi(G)$ is defined as the minimum n for which G has an n-coloring. A graph G is *n-colorable* if $\chi(G) \leq n$ and is *n-chromatic* if $\chi(G) = n$.

Since G obviously has a p-coloring and a $\chi(G)$-coloring, it must also have an n-coloring whenever $\chi(G) < n < p$. The graph of Fig. 12.1 is 2-chromatic; n-colorings for $n = 2, 3, 4$ are displayed, with positive integers designating the colors.

The chromatic numbers of some of the familiar graphs are easily determined, namely $\chi(K_p) = p$, $\chi(K_p - x) = p - 1$, $\chi(\bar{K}_p) = 1$, $\chi(K_{m,n}) = 2$, $\chi(C_{2n}) = 2$, $\chi(C_{2n+1}) = 3$, and for any nontrivial tree T, $\chi(T) = 2$.

Obviously, a graph is 1-chromatic if and only if it is totally disconnected. A characterization of bicolorable (2-colorable) graphs was given by König [K10, p. 170], as Theorem 2.4 already indicates.

Theorem 12.1 A graph is bicolorable if and only if it has no odd cycles.

It is likely to remain an unsolved problem to provide a characterization of n-colorable graphs for $n \geq 3$, since such a criterion even for $n = 3$ would help to settle the 4CC. No convenient method is known for determining the chromatic number of an arbitrary graph. However, there are several known bounds for $\chi(G)$ in terms of various other invariants. One obvious lower bound is the number of points in a largest complete subgraph of G. We now consider upper bounds, the first of which is due to Szekeres and Wilf [SW1].

Theorem 12.2 For any graph G,

$$\chi(G) \leq 1 + \max \delta(G'),\tag{12.1}$$

where the maximum is taken over all induced subgraphs G' of G.

Proof. The result is obvious for totally disconnected graphs. Let G be an arbitrary n-chromatic graph, $n \geq 2$. Let H be any smallest induced subgraph such that $\chi(H) = n$. The graph H therefore has the property that $\chi(H - v) = n - 1$ for all its points v. It follows that deg $v \geq n - 1$ so that $\delta(H) \geq n - 1$ and hence

$$n - 1 \leq \delta(H) \leq \max \delta(H') \leq \max \delta(G'),$$

the first maximum taken over all induced subgraphs H' of H and the second over all induced subgraphs G' of G. This implies that

$$\chi(G) = n \leq 1 + \max \delta(G').$$

Corollary 12.2(a) For any graph G, the chromatic number is at most one greater than the maximum degree,

$$\chi \leq 1 + \Delta. \tag{12.2}$$

Brooks [B16] showed, however, that this bound can often be improved.

Theorem 12.3 If $\Delta(G) = n \geq 2$, then G is n-colorable unless

i) $n = 2$ and G has a component which is an odd cycle, or

ii) $n > 2$ and K_{n+1} is a component of G.

A lower bound, noted in Berge [B12, p. 37] and Ore [O5, p. 225], and an upper bound, Harary and Hedetniemi [HH1], involve the point independence number β_0 of G.

Theorem 12.4 For any graph G,

$$p/\beta_0 \leq \chi \leq p - \beta_0 + 1. \tag{12.3}$$

Proof. If $\chi(G) = n$, then V can be partitioned into n color classes $V_1, V_2, \cdots,$ V_n, each of which, as noted above, is an independent set of points. If $|V_i| = p_i$, then every $p_i \leq \beta_0$ so that $p = \Sigma\, p_i \leq n\beta_0$.

To verify the upper bound, let S be a maximal independent set containing β_0 points. It is clear that $\chi(G - S) \geq \chi(G) - 1$. Since $G - S$ has $p - \beta_0$ points, $\chi(G - S) \leq p - \beta_0$. Therefore, $\chi(G) \leq \chi(G - S) + 1 \leq p - \beta_0 + 1$.

None of the bounds presented here is particularly good in the sense that for any bound and for every positive integer n, there exists a graph G such that $\chi(G)$ differs from the bound by more than n.

From the discussion thus far, one may very well be led to believe that all graphs with large chromatic number have large cliques and hence contain triangles. In fact, Dirac [D7] asked if there exists a graph with no triangles but arbitrarily high chromatic number. This was answered affirmatively and independently by Blanche Descartes* [D3], Mycielski [M19], and Zykov [Z1]. Their result was extended by Kelly and Kelly [KK1], who proved that for all $n \geq 2$, there exists an n-chromatic graph whose girth exceeds 5. In the same paper, they conjectured the following theorem, which was first proved by Erdös [E2] using a probabilistic argument and later by Lovász [L5] constructively.

---

* This so-called lady is actually a nonempty subset of {Brooks, Smith, Stone, Tutte}; in this case {Tutte}.

Theorem 12.5 For every two positive integers m and n, there exists an n-chromatic graph whose girth exceeds m.

The number $\bar{\chi} = \bar{\chi}(G) = \chi(\bar{G})$ is the minimum number of subsets which partition the point set of G so that each subset induces a complete subgraph of G. It is clear that $\bar{\chi}(G) \geq \beta_0(G)$. Bounds on the sum and product of the chromatic numbers of a graph and its complement were developed by Nordhaus and Gaddum [NG1].

Theorem 12.6 For any graph G, the sum and product of χ and $\bar{\chi}$ satisfy the inequalities:

$$2\sqrt{p} \leq \chi + \bar{\chi} \leq p + 1, \tag{12.4}$$

$$p \leq \chi\bar{\chi} \leq \left(\frac{p+1}{2}\right)^2. \tag{12.5}$$

Proof. Let G be n-chromatic and let $V_1, V_2, \cdots, V_n$ be the color classes of G, where $|V_i| = p_i$. Then of course $\Sigma\, p_i = p$ and max $p_i \geq p/n$. Since each V_i induces a complete subgraph of $\bar{G}$, $\bar{\chi} \geq$ max $p_i \geq p/n$ so that $\chi\bar{\chi} \geq p$. Since the geometric mean of two positive numbers never exceeds their arithmetic mean, it follows that $\chi + \bar{\chi} \geq 2\sqrt{p}$. This establishes both lower bounds.

To show that $\chi + \bar{\chi} \leq p + 1$, we use induction on p, noting that equality holds when $p = 1$. We thus assume that $\chi(G) + \bar{\chi}(G) \leq p$ for all graphs G having $p - 1$ points. Let H and $\bar{H}$ be complementary graphs with p points, and let v be a point of H. Then $G = H - v$ and $\bar{G} = \bar{H} - v$ are complementary graphs with $p - 1$ points. Let the degree of v in H be d so that the degree of v in $\bar{H}$ is $p - d - 1$. It is obvious that

$$\chi(H) \leq \chi(G) + 1 \qquad \text{and} \qquad \bar{\chi}(H) \leq \bar{\chi}(G) + 1.$$

If either

$$\chi(H) < \chi(G) + 1 \qquad \text{or} \qquad \bar{\chi}(H) < \bar{\chi}(G) + 1,$$

then $\chi(H) + \bar{\chi}(H) \leq p + 1$. Suppose then that $\chi(H) = \chi(G) + 1$ and $\bar{\chi}(H) = \bar{\chi}(G) + 1$. This implies that the removal of v from H, producing G, decreases the chromatic number so that $d \geq \chi(G)$. Similarly

$$p - d - 1 \geq \bar{\chi}(G);$$

thus $\chi(G) + \bar{\chi}(G) \leq p - 1$. Therefore, we always have

$$\chi(H) + \bar{\chi}(H) \leq p + 1.$$

Finally, applying the inequality $4\chi\bar{\chi} \leq (\chi + \bar{\chi})^2$ we see that

$$\chi\bar{\chi} \leq [(p + 1)/2]^2.$$

THE FIVE COLOR THEOREM

Although it is not known whether all planar graphs are 4-colorable, they are certainly 5-colorable. In this section we present a proof of this famous result due to Heawood [H38].

Theorem 12.7 Every planar graph is 5-colorable.

Proof. We proceed by induction on the number p of points. For any planar graph having $p \leq 5$ points, the result follows trivially since the graph is p-colorable.

As the inductive hypothesis we assume that all planar graphs with p points, $p \geq 5$, are 5-colorable. Let G be a plane graph with $p + 1$ vertices. By Corollary 11.1(e), G contains a vertex v of degree 5 or less. By hypothesis, the plane graph $G - v$ is 5-colorable.

Consider an assignment of colors to the vertices of $G - v$ so that a 5-coloring results, where the colors are denoted by c_i, $1 \leq i \leq 5$. Certainly, if some color, say c_j, is not used in the coloring of the vertices adjacent with v, then by assigning the color c_j to v, a 5-coloring of G results.

This leaves only the case to consider in which deg $v = 5$ and five colors are used for the vertices of G adjacent with v. Permute the colors, if necessary, so that the vertices colored c_1, c_2, c_3, c_4, and c_5 are arranged cyclically about v. Now label the vertex adjacent with v and colored c_i by v_i, $1 \leq i \leq 5$ (see Fig. 12.2).

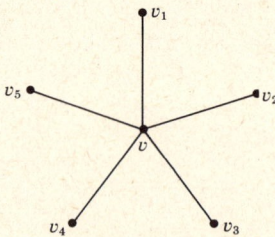

Fig. 12.2. A step in the proof of the Five Color Theorem.

Let G_{13} denote the subgraph of $G - v$ induced by those vertices colored c_1 or c_3. If v_1 and v_3 belong to different components of G_{13}, then a 5-coloring of $G - v$ may be accomplished by interchanging the colors of the vertices in the component of G_{13} containing v_1. In this 5-coloring, however, no vertex adjacent with v is colored c_1, so by coloring v with the color c_1, a 5-coloring of G results.

If, on the other hand, v_1 and v_3 belong to the same component of G_{13}, then there exists in G a path between v_1 and v_3 all of whose vertices are colored c_1 or c_3. This path together with the path $v_1 v v_3$ produces a cycle which necessarily encloses the vertex v_2 or both the vertices v_4 and v_5. In any case, there exists no path joining v_2 and v_4, all of whose vertices are colored c_2 or c_4. Hence, if we let G_{24} denote the subgraph of $G - v$ induced by the vertices colored c_2 or c_4, then v_2 and v_4 belong to different components of G_{24}. Thus if we interchange colors of the vertices in the component of G_{24} containing v_2, a 5-coloring of $G - v$ is produced in which no vertex adjacent with v is colored c_2. We may then obtain a 5-coloring of G by assigning to v the color c_2.

THE FOUR COLOR CONJECTURE

In Chapter 1 we mentioned that the 4CC served as a catalyst for graph theory through attempts to settle it. We now present a graph-theoretic discussion of this infamous problem. A *coloring of a plane map* G is an assignment of colors to the regions of G so that no two adjacent regions are assigned the same color. The map G is said to be *n-colorable* if there is a coloring of G which uses n or fewer colors. The original conjecture as described in Chapter 1 asserts that every plane map is 4-colorable.

Four Color Conjecture (4CC)   Every planar graph is 4-colorable.

We emphasize that coloring a graph always refers to coloring its vertices while coloring a map indicates that it is the regions which are colored! Thus the conjecture that every plane map is 4-colorable is in fact equivalent to this statement of the Four Color Conjecture. To see this, assume the 4CC holds and let G be any plane map. Let G^* be the underlying graph of the geometric dual of G. Since two regions of G are adjacent if and only if the corresponding vertices of G^* are adjacent, map G is 4-colorable because graph G^* is 4-colorable.

Conversely, assume that every plane map is 4-colorable and let H be any planar graph. Without loss of generality, we suppose H is a connected plane graph. Let H^* be the dual of H, so drawn that each region of H^* encloses precisely one vertex of H. The connected plane pseudograph H^* can be converted into a plane graph H' by introducing two vertices into each loop of H^* and adding a new vertex into each edge in a set of multiple edges. The 4-colorability of H' now implies that H is 4-colorable, completing the verification of the equivalence.

If the 4CC is ever proved, the result will be best possible, for it is easy to give examples of planar graphs which are 4-chromatic, such as K_4 and W_6 (see Fig. 12.3).

Each of the graphs K_4 and W_6 has more than 3 triangles, which is necessary according to a theorem of Grünbaum [G9].

Theorem 12.8   Every planar graph with fewer than 4 triangles is 3-colorable.

From this the following corollary is immediate; it was originally proved by Grötzsch [G8].

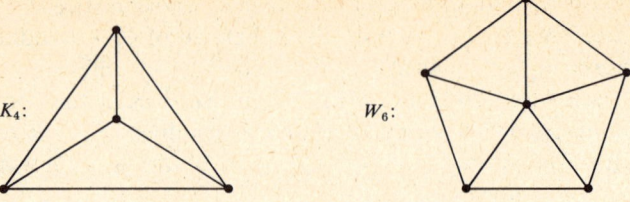

Fig. 12.3. Two 4-chromatic planar graphs.

Corollary 12.8(a) Every planar graph without triangles is 3-colorable.

Any plane map which requires 5 colors will necessarily contain a large number of regions, for Ore and Stemple [OS1] showed that all plane maps with up to 39 regions are 4-colorable, increasing by 4 regions the earlier result of this kind.* All evidence indicates that the Four Color Conjecture is true. However, attempts to prove the 4CC using the plane map formulation can be directed at a special class of plane maps, as we shall now see.

Theorem 12.9 The Four Color Conjecture holds if and only if every cubic bridgeless plane map is 4-colorable.

Proof. We have already seen that every plane map is 4-colorable if and only if the 4CC holds. This is also equivalent to the statement that every bridgeless plane map is 4-colorable since the elementary contraction of identifying the endvertices of a bridge affects neither the number of regions in the map nor the adjacency of any of the regions.

Certainly, if every bridgeless plane map is 4-colorable, then every cubic bridgeless plane map is 4-colorable. In order to verify the converse, let G be a bridgeless plane map and assume all cubic bridgeless plane maps are 4-colorable. Since G is bridgeless, it has no endvertices. If G contains a vertex v of degree 2 incident with edges y and z, we subdivide y and z, denoting the subdivision vertices by u and w, respectively. We now remove v, identify u with one of the vertices of degree 2 in a copy of the graph $K_4 - x$ and identify w with the other vertex of degree 2 in $K_4 - x$. Observe that each new vertex added has degree 3 (see Fig. 12.4). If G contains a vertex v_0 of degree $n \geq 4$ incident with edges $x_1, x_2, \cdots, x_n$, arranged cyclically about v_0, we subdivide each x_i producing a new vertex v_i. We then remove v_0 and add the new edges $v_1 v_2, v_2 v_3, \cdots, v_{n-1} v_n, v_n v_1$. Again each of the vertices so added has degree 3.

Denote the resulting bridgeless cubic plane map by G', which, by hypothesis, is 4-colorable. If for each vertex v of G with deg $v \neq 3$, we

---

* Finck and Sachs [FS1] proved that every plane graph with at most 21 triangles is 4-colorable.

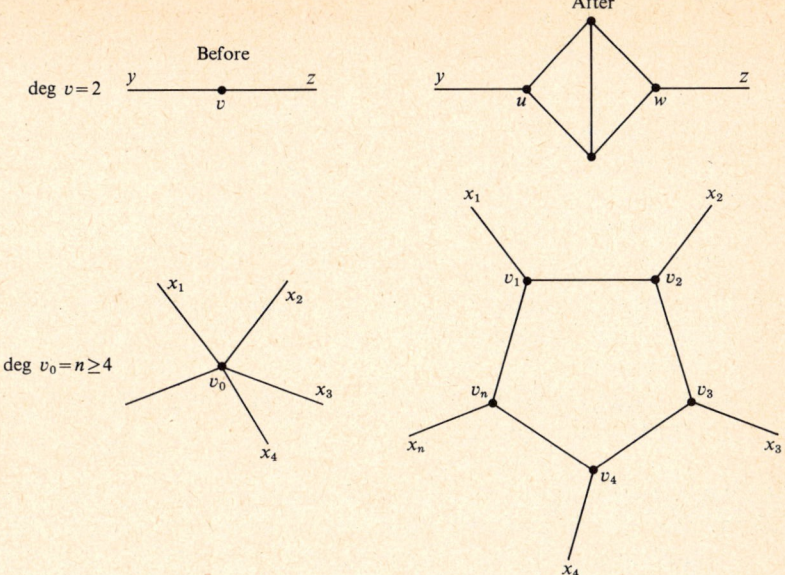

Fig. 12.4. Conversion of a graph into a cubic graph.

identify all the newly added vertices associated with v in the formation of G', we arrive at G once again. Thus let there be given a 4-coloring of G'. The aforementioned contraction of G' into G induces an m-coloring of G, $m \leq 4$, which completes the proof.

Another interesting equivalence was proved by Whitney [W16].

Theorem 12.10   The Four Color Conjecture holds if and only if every hamiltonian planar graph is 4-colorable.

As there are equivalents of the Four Color Conjecture involving the coloring of regions, so too is there an equivalent of the 4CC concerned with the coloring of lines.

A *line-coloring* of a graph G is an assignment of colors to its lines so that no two adjacent lines are assigned the same color. An *n-line-coloring* of G is a line-coloring of G which uses exactly n colors. The *line-chromatic number** $\chi'(G)$ is the minimum n for which G has an n-line-coloring. It follows that for any graph G which is not totally disconnected, $\chi'(G) = \chi(L(G))$. Tight bounds on the line-chromatic number were obtained** by Vizing [V4].

Theorem 12.11   For any graph G, the line-chromatic number satisfies the inequalities:

$$\Delta \leq \chi' \leq \Delta + 1. \tag{12.6}$$

---

* Sometimes called the chromatic index.
** A proof in English can be found in Ore [O7, p. 248].

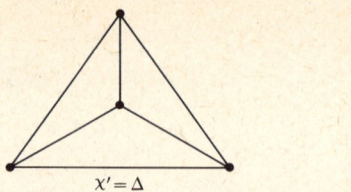
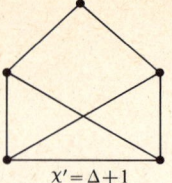

$\chi' = \Delta$ $\chi' = \Delta + 1$

Fig. 12.5. The two values for the line-chromatic number.

The two possible values for $\chi'(G)$ are illustrated in Fig. 12.5. It is not known in general for which graphs $\chi' = \Delta$.

Theorem 12.12 The Four Color Conjecture is true if and only if $\chi'(G) = 3$ for every bridgeless cubic planar graph G.

Proof. We have already shown in Theorem 12.9 that the 4CC is equivalent to the statement that every cubic bridgeless plane map is 4-colorable. We show now that a cubic bridgeless plane map G is 4-colorable if and only if $\chi'(G) = 3$.

First we assume that G is a bridgeless, cubic plane map which is 4-colorable. Without loss of generality, we take G to be connected and therefore a plane map which, by hypothesis, is 4-colorable. For the set of colors we select the elements of the Klein four-group F, where addition in F is defined by $k_i + k_i = k_0$ and $k_1 + k_2 = k_3$, with k_0 the identity element.

Let there be given a 4-coloring of the map G. We define the color of an edge to be the sum of the colors of the two distinct regions which are incident with the edge. It is now immediate that the edges are colored with elements of the set $\{k_1, k_2, k_3\}$ and that no two adjacent edges are assigned the same color; thus $\chi'(G) = 3$.

Conversely, let G be a bridgeless cubic plane graph with $\chi'(G) = 3$, and color its edges with the three nonzero elements of F. Select some region R_0 and assign to it the color k_0. To any other region R of G, we assign a color in the following manner. Let C be any curve in the plane joining the interior of R_0 with the interior of R such that C does not pass through a point of G. We then define the color of R to be the sum of the colors of those edges which intersect C.

That the colors of the regions are well-defined depends on the fact that the sum of the colors of the edges which intersect any simple closed curve not passing through a vertex of G is k_0. Let S be such a curve, and let $c_1, c_2, \cdots, c_n$ be the colors of the edges which intersect S. In addition, let $d_1, d_2, \cdots, d_m$ be the colors of those edges interior to S. Observe that if $c(v)$ denotes the sum of the colors of the 3 edges incident with a vertex v, then $c(v) = k_0$. Hence for all vertices v interior to S, $\Sigma\, c(v) = k_0$. On the other hand, we also have

$$\sum c(v) = c_1 + c_2 + \cdots + c_n + 2(d_1 + d_2 + \cdots + d_m)$$
$$= c_1 + c_2 + \cdots + c_n$$

since every element of F is self-inverse. Thus $c_1 + c_2 + \cdots + c_n = k_0$. It is now a routine matter to show that this constitutes a 4-coloring of the regions of G, completing the proof.

Since each line color class resulting from an n-line coloring of a regular graph G of degree n is a 1-factor of G, the preceding result produces another equivalent of the Four Color Conjecture.

Corollary 12.12(a) The Four Color Conjecture holds if and only if every bridgeless, cubic planar graph is 1-factorable.

Theorem 12.12 has been generalized in terms of factorization (see Ore [O7, p. 103]).

Theorem 12.13 A necessary and sufficient condition that a connected planar map G be 4-colorable is that G be the sum of three subgraphs G_1, G_2, G_3 such that for each point v, the number of lines of each G_i incident with v are all even or all odd.

Although it is the 4CC which has received the preponderance of publicity, there are several other conjectures dealing with coloring. One of the most interesting of these involves contractions and is due to Hadwiger [H1].

Hadwiger's Conjecture. Every connected n-chromatic graph is contractible to K_n.

Not surprisingly, this conjecture is related to the 4CC. Hadwiger's Conjecture is known to be true for $n \leq 4$, a result of Dirac [D5]. For $n = 5$, this conjecture states that every 5-chromatic graph G is contractible to K_5. By Theorem 11.14, every such graph G is necessarily nonplanar. Thus Hadwiger's Conjecture for $n = 5$ implies the 4CC. The converse was established by Wagner [W3].

Theorem 12.14 Hadwiger's Conjecture for $n = 5$ is equivalent to the Four Color Conjecture.

THE HEAWOOD MAP-COLORING THEOREM

Let S_n be the orientable surface of genus n; thus, S_n is topologically equivalent to a sphere with n handles. The *chromatic number of S_n*, denoted $\chi(S_n)$, is the maximum chromatic number among all graphs which can be embedded on S_n. The surface S_0 is simply the sphere and the determination of $\chi(S_0)$ is the problem we have already encountered on several occasions. The Four

Color Conjecture states that $\chi(S_0) = 4$ although, of course, we know only (by Theorem 12.7) that $\chi(S_0)$ is 4 or 5.

For the torus, Heawood [H38] was able to prove that $\chi(S_1) = 7$. The inequality $\chi(S_1) \geq 7$ follows from the fact that it is possible to embed K_7 on the torus. This is shown in Fig. 11.17. The equality $\chi(S_1) = 7$ comes from the fact that Heawood was also able to prove (see the proof of Theorem 12.15 below) that the chromatic number of the orientable surface of positive genus n has the upper bound

$$\chi(S_n) \leq \left[\frac{7 + \sqrt{1 + 48n}}{2} \right] \qquad (n > 0). \qquad (12.7)$$

For $n = 1$, we have $\chi(S_1) \leq 7$, so that $\chi(S_1) = 7$.

Heawood, who found the error in Kempe's "proof" of the Four Color Conjecture, was himself not infallible. He believed that he had proved equality in his formula, but just one year later, Heffter [H40] pointed out errors of omission in Heawood's arguments resulting in only the inequality (12.7). Heffter did prove equality for $0 < n \leq 6$. Eventually, the statement that equality holds in Heawood's formula became known as the Heawood Map-Coloring Conjecture. We now show that when Ringel and Youngs proved that $\gamma(K_p) = \{(p - 3)(p - 4)/12\}$, Theorem 11.18, they settled this conjecture.

Theorem 12.15 (Heawood Map-Coloring Theorem). For every positive integer n, the chromatic number of the orientable surface of genus n is given by

$$\chi(S_n) = \left[\frac{7 + \sqrt{1 + 48n}}{2} \right] \qquad (n > 0). \qquad (12.8)$$

Proof. We first prove inequality (12.7). Let G be a (p, q) graph embedded on S_n. We may assume G is a triangulation, since any graph can be augmented to a triangulation of the same genus by adding edges, without reducing χ. If $\bar{d}$ is the average degree of the vertices of G, then p, q, and r (the number of regions) are related by the equations

$$\bar{d}p = 2q = 3r. \qquad (12.9)$$

Solving for q and r in terms of p and using Euler's equation (11.4), we obtain

$$\bar{d} = 12(n - 1)/p + 6. \qquad (12.10)$$

Since $\bar{d} \leq p - 1$, this gives the inequality

$$p - 1 \geq 12(n - 1)/p + 6. \qquad (12.11)$$

Solving for p and taking the positive root, we obtain

$$p \geq \left[\frac{7 + \sqrt{1 + 48n}}{2} \right]. \qquad (12.12)$$

Let $H(n)$ be the right-hand side of (12.8). Then we must show that $H(n)$ colors are sufficient to color the points of G. Clearly if $p = H(n)$ we have enough colors. If, on the other hand, $p > H(n)$, we substitute $H(n)$ for p in (12.10), to obtain the inequality

$$\bar{d} < 12(n - 1)/H(n) + 6 = H(n) - 1, \qquad (12.13)$$

with the latter equality obtained by routine algebraic manipulation. Thus when $p > H(n)$, there is a point v of degree at most $H(n) - 2$. Identify v and any adjacent point (by an elementary contraction) to obtain a new graph G'. If $p' = p - 1 = H(n)$, then G' can be colored in $H(n)$ colors. If $p' > H(n)$, repeat the argument. Eventually an $H(n)$-colorable graph will be obtained. It is then easy to see that the coloring of this graph induces a coloring of the preceding one in $H(n)$ colors, and so forth, so that G itself is $H(n)$-colorable.

The other half of the theorem is the difficult part, but Ringel and Youngs have provided the means. If the complete graph K_p can be embedded in S_n, then by equation (11.9),

$$n \geq \gamma(K_p) = \left\{ \frac{(p - 3)(p - 4)}{12} \right\}. \qquad (12.14)$$

Setting p to be the largest integer satisfying Eq. (12.14), we have

$$\frac{(p - 3)(p - 4)}{12} \leq \left\{ \frac{(p - 3)(p - 4)}{12} \right\} \leq n \leq \left\{ \frac{(p - 2)(p - 3)}{12} \right\} - 1 < \frac{(p - 2)(p - 3)}{12}.$$

Solving for p we find

$$\frac{5 + \sqrt{1 + 48n}}{2} \leq p \leq \frac{7 + \sqrt{1 + 48n}}{2}.$$

So

$$p = \left[\frac{7 + \sqrt{1 + 48n}}{2} \right]. \qquad (12.15)$$

Since $\chi(K_p) = p$, we have found a graph with genus n and chromatic number equal to $H(n)$. This shows that $H(n)$ is a lower bound for $\chi(S_n)$ and completes the proof. *Note:* (12.8) specialized to $n = 0$ is precisely the 4CC.

UNIQUELY COLORABLE GRAPHS

Let G be a labeled graph. Any $\chi(G)$-coloring of G induces a partition of the point set of G into $\chi(G)$ color classes. If $\chi(G) = n$ and every n-coloring of G induces the same partition of V, then G is called *uniquely n-colorable* or simply *uniquely colorable*. The graph G of Fig. 12.6 is uniquely 3-colorable since every 3-coloring of G has the partition $\{u_1\}, \{u_2, u_4\}, \{u_3, u_5\}$ while the pentagon is not uniquely 3-colorable; indeed, five different partitions of its point set are possible.

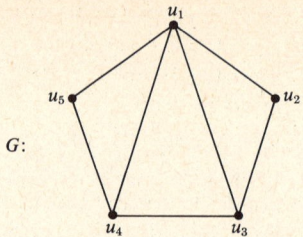

Fig. 12.6. A uniquely colorable graph.

We begin with a few elementary observations concerning uniquely colorable graphs. First, in any n-coloring of a uniquely n-colorable graph G, every point v of G is adjacent with at least one point of every color different from that assigned to v; for otherwise a different n-coloring of G could be obtained by recoloring v. This further implies that $\delta(G) \geq n - 1$. A necessary condition for a graph to be uniquely colorable was found by Cartwright and Harary [CH2].

Theorem 12.16 In the n-coloring of a uniquely n-colorable graph, the subgraph induced by the union of any two color classes is connected.

Proof. Consider an n-coloring of a uniquely n-colorable graph G, and suppose there exist two color classes of G, say C_1 and C_2, such that the subgraph S of G induced by $C_1 \cup C_2$ is disconnected. Let S_1 and S_2 be two components of S. From our earlier remarks, each of S_1 and S_2 must contain points of both C_1 and C_2. An n-coloring different from the given one can now be obtained if the color of the points in $C_1 \cap S_1$ is interchanged with the color of the points in $C_2 \cap S_1$. This implies that G is not uniquely n-colorable, which is a contradiction.

The converse of Theorem 12.16 is not true, however. This can be seen with the aid of the 3-chromatic graph G of Fig. 12.7. It has the property that in any 3-coloring, the subgraph induced by the union of any 2 color classes is connected, but G is not uniquely 3-colorable.

From Theorem 12.16, it now follows that every uniquely n-colorable graph, $n \geq 2$, is connected. However, a stronger result can be given, due to Chartrand and Geller [CG1].

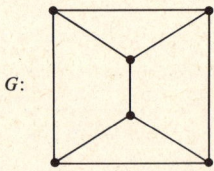

Fig. 12.7. A counterexample to the converse of Theorem 12.16.

that G is neither complete nor $(n - 1)$-connected so that there exists a set U of $n - 2$ points whose removal disconnects G. Thus, there are at least two distinct colors, say c_1 and c_2, not assigned to any point of U. By Theorem 12.16, a point colored c_1 is connected to any point colored c_2 by a path all of whose points are colored c_1 or c_2. Hence, the set of points of G colored c_1 or c_2 lies within the same component of $G - U$, say G_1. Another n-coloring of G can therefore be obtained by taking any point of $G - U$ which is not in G_1 and recoloring it either c_1 or c_2. This contradicts the hypothesis that G is uniquely n-colorable; thus G is $(n - 1)$-connected.

Since the union of any k color classes of a uniquely n-colorable graph, $2 \le k \le n$, induces a uniquely k-colorable graph, we arrive at the following consequence.

Corollary 12.17(a) In any n-coloring of a uniquely n-colorable graph, the subgraph induced by the union of any k color classes, $2 \le k \le n$, is $(k - 1)$-connected.

It is easy to give examples of 3-chromatic graphs containing no triangles; indeed we have seen in Theorem 12.5 that for any n, there exist n-chromatic graphs with no triangles and hence no subgraphs isomorphic to K_n. In this connection, a stronger result was obtained by Harary, Hedetniemi, and Robinson [HHR1].

Theorem 12.18 For all $n \ge 3$, there is a uniquely n-colorable graph which contains no subgraph isomorphic to K_n.

For $n = 3$, the graph G of Fig. 12.8 illustrates the theorem.

Naturally, a graph is uniquely 1-colorable if and only if it is 1-colorable, that is, totally disconnected. It is also well known that a graph G is uniquely 2-colorable if and only if G is 2-chromatic and connected. As might be expected, the information concerning uniquely n-colorable graphs, $n \ge 3$, is

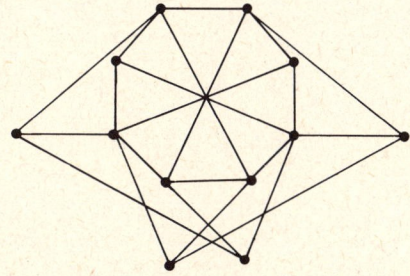

Fig. 12.8. A uniquely 3-colorable graph having no triangles.

very sparse. In the case where the graphs are planar, however, more can be said, although in view of the Five Color Theorem, we need to consider only the values $3 \le n \le 5$. The results in this area are due to Chartrand and Geller [CG1].

Theorem 12.19 Let G be a 3-chromatic plane graph. If G contains a triangle T such that for each vertex v of G there is a sequence $T, T_1, T_2, \cdots, T_m$ of triangles with v in T_m such that consecutive triangles in the sequence have an edge in common, then G is uniquely 3-colorable.

The next result is now immediate.

Corollary 12.19(a) If a 2-connected 3-chromatic plane graph G has at most one region which is not a triangle, then G is uniquely 3-colorable.

The converse of Corollary 12.19(a) is not true, for a uniquely 3-colorable planar graph may have more than one region which is not a triangle; see Fig. 12.9. However, every uniquely 3-colorable planar graph must contain triangles.

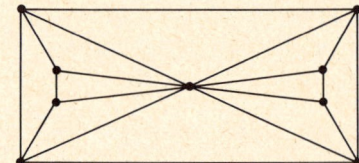

Fig. 12.9. A uniquely 3-colorable planar graph.

Theorem 12.20 If G is a uniquely 3-colorable planar graph with at least 4 points, then G contains at least two triangles.

In the case of uniquely 4-colorable planar graphs, the situation is particularly simple.

Theorem 12.21 Every uniquely 4-colorable planar graph is maximal planar.

Proof. Let there be given a 4-coloring of a uniquely 4-colorable planar graph G with the color classes denoted by V_i, $1 \le i \le 4$, where $|V_i| = p_i$. Since the subgraph induced by $V_i \cup V_j$, $i \ne j$, is connected, G must have at least $\Sigma(p_i + p_j - 1)$ lines, $1 \le i < j \le 4$. However, this sum is obviously $3p - 6$. Hence $q \ge 3p - 6$ and so by Corollary 11.1(b), G is maximal planar.

Although the existence of a 5-chromatic planar graph is still open, a result of Hedetniemi given in [CG1] settles the problem for unique 5-colorability; its proof is similar to that of the preceding theorem.

Theorem 12.22 No planar graph is uniquely 5-colorable.

CRITICAL GRAPHS

If the Four Color Conjecture is not true, then there must exist a smallest 5-chromatic planar graph. Such a graph G has the property that for every point v, the subgraph $G - v$ is 4-chromatic. Thus we have a natural approach to a possible proof of the 4CC in its contrapositive formulation. This suggests the basic problem of investigating such 5-chromatic graphs G or, more generally, those n-chromatic graphs G with the property that $\chi(G - v) = n - 1$ for all points v of G.

Following Dirac [D5], a graph G is called *critical** if $\chi(G - v) < \chi(G)$ for all points v; if $\chi(G) = n$, then G is n-critical. Of course, if G is critical, then $\chi(G - v) = \chi(G) - 1$ for every point v.

Obviously, no graph is 1-critical. The only 2-critical graph is K_2, while the only 3-critical graphs are the odd cycles. For $n \geq 4$, the n-critical graphs have not been characterized.

Ordinarily, it is extremely difficult to determine whether a given graph is critical; however, every n-chromatic graph, $n \geq 2$, contains an n-critical subgraph. In fact, if H is any smallest induced subgraph of G such that $\chi(H) = \chi(G)$, then H is critical.

It is clear that every critical graph G is connected; furthermore, since $\chi(G) = \max \chi(B)$ over all blocks B of G, it follows that G must be a block. This is only one of several properties which critical graphs enjoy.

The next statement has already been demonstrated within the proof of Theorem 12.2.

Theorem 12.23 If G is an n-critical graph, then $\delta(G) \geq n - 1$.

We now make an observation on the removal of points.

Theorem 12.24 No critical graph can be separated by a complete subgraph.

Corollary 12.24(a) Every cutset of points of a critical graph contains two nonadjacent points.

Every complete graph is critical; indeed for $U \subset V(K_p)$, $\chi(K_p - U) = p - |U|$. For any other critical graph, however, it is always possible to remove more than one point without decreasing the chromatic number by more than one; in fact, if S is any independent set of points of an n-critical graph, then $\chi(G - S) = n - 1$. This further implies that if u and v are any two nonadjacent points of an n-critical graph G which is not complete, there exists an n-coloring of G such that u and v are in the same color class and an n-coloring of G such that u and v are in different color classes.

---

* If other kinds of critical graphs are present, these should be called color-critical.

One area of research on critical graphs deals with cycle length, in particular with circumference and girth. By Theorem 12.23 and Corollary 7.3(b), if G is an n-critical graph with p points such that $p \leq 2n - 2$, then G is hamiltonian. More generally, Dirac [D6] proved the following result.

Theorem 12.25 If G is an n-critical graph, $n \geq 3$, then either G is hamiltonian or the circumference of G is at least $2n - 2$.

Dirac [D6] once conjectured that every 4-critical graph is hamiltonian; however, Kelly and Kelly [KK1] showed this conjecture is not true. Dirac [D6] also conjectured that for all m and n, $n \geq 3$, there exists a sufficiently large value of p such that all n-critical graphs with at least p points have circumference exceeding m. Kelly and Kelly proved this to be true. It is a consequence of Theorem 12.5 that for all m and n, there exists an n-critical graph whose girth exceeds m.

A critical graph G may have the added property that for any line x of G, $\chi(G - x) = \chi(G) - 1$; in such a case, G is called *line-critical*, and if $\chi(G) = n$, G is *n-line-critical*. Although every line-critical graph without isolated points is necessarily critical, the converse does not hold. For example, the graph G of Fig. 12.10 is 4-critical but is not line-critical since $\chi(G - x) = 4$.

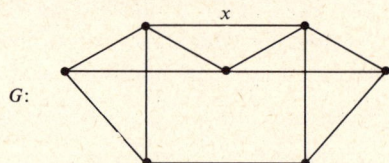

Fig. 12.10. A critical graph which is not line-critical.

Thus every property of critical graphs is also possessed by line-critical graphs; but in some instances more can be said about the latter.

Theorem 12.26 If G is a 2-connected n-chromatic graph containing exactly one point of degree exceeding $n - 1$, then G is n-line-critical.

Proof. Let x be any line of G, and consider $G - x$. Certainly, $\delta(G - x) \leq n - 2$, and, moreover, for every induced subgraph G' of $G - x$, $\delta(G') \leq n - 2$. Thus by Theorem 12.2, $\chi(G - x) \leq n - 1$, implying that $\chi(G - x) = n - 1$ and that G is n-line-critical.

According to Theorem 12.23, if G is an n-critical graph, then $2q \geq (n - 1)p$. For line-critical graphs, however, Dirac [D7] improved this result.

Theorem 12.27 If G is an n-line-critical graph without isolated points, $n \geq 4$, which is not complete, then

$$2q \geq (n - 1)p + n - 3.$$

HOMOMORPHISMS

It is convenient to consider only connected graphs in this section. An *elementary homomorphism* of G is an identification of two nonadjacent points. A *homomorphism* of G is a sequence of elementary homomorphisms. If G' is the graph resulting from a homomorphism ϕ of G we can consider ϕ as a function from V onto V' such that if u and v are adjacent in G, then ϕu and ϕv are adjacent in G'. Note that every line of G' must come from some line of G, that is, if u' and v' are adjacent in G', then there are two adjacent points u and v in G such that $\phi u = u'$ and $\phi v = v'$. We say that ϕ is a *homomorphism* of G onto G', that G' is a *homomorphic image* of G, and write $G' = \phi G$. Thus in particular every isomorphism is a homomorphism. The path P_4 has just 4 homomorphic images, shown in Fig. 12.11.

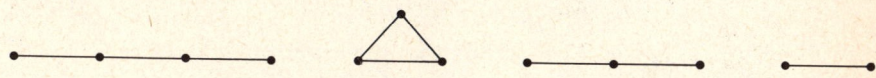

Fig. 12.11.   The homomorphic images of path P_4.

A homomorphism ϕ of G is *complete of order n* if $\phi G = K_n$. Note that any homomorphism ϕ of G onto K_n corresponds to an n-coloring of G since the points of K_n can be regarded as colors and by definition of homomorphism no two points of G with the same color are adjacent. Each coloring defined by a complete homomorphism has the property that for any two colors, there are adjacent points u and v of G colored with these colors. In this case we have a *complete coloring*. Figure 12.12 shows a graph with complete colorings of order 3 and 4, where colors are indicated by positive integers. Obviously the smallest order of all complete homomorphisms of G must be $\chi(G)$.

The next theorem [HHP1] generalizes an earlier result due to Hajós [H3] which appears as its corollary.

Theorem 12.28 For any graph G and any elementary homomorphism ε of G,

$$\chi(G) \leq \chi(\varepsilon G) \leq 1 + \chi(G). \tag{12.16}$$

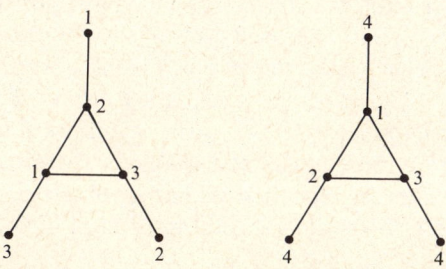

Fig. 12.12.   Two complete colorings of a graph.

Proof. Let ε be the elementary homomorphism of G which identifies the nonadjacent points u and v. Then any coloring of εG yields a coloring of G when the same color is used for u and v, so $\chi(G) \leq \chi(\varepsilon G)$. On the other hand, a coloring of εG is obtained from a coloring of G when the new point is given a color different from all those used in coloring G, so that $\chi(\varepsilon G) \leq 1 + \chi(G)$.

Corollary 12.28(a) For any homomorphism ϕ of G, $\chi(G) \leq \chi(\phi G)$.

It is now natural to consider the maximum order of all complete homomorphisms of G. This invariant is called the *achromatic number* and is denoted $\psi(G)$. Since at most p colors can be used, it is obvious that $\chi(G) \leq \psi(G) \leq p$. Neither of these inequalities is a particularly good bound for ψ.

Theorem 12.29 For any graph G and any elementary homomorphism ε of G,

$$\psi(G) - 2 \leq \psi(\varepsilon G) \leq \psi(G). \tag{12.17}$$

The example in Fig. 12.13 shows that the lower bound can be attained, and hence is best possible. It is easy to verify that $\psi(G) = 5$ while $\psi(\varepsilon G) = 3$.

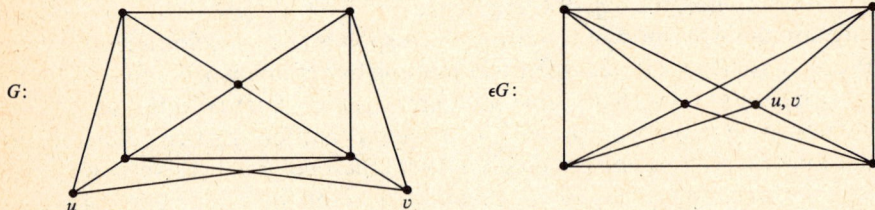

Fig. 12.13. A homomorphism which decreases ψ by 2.

The next result, called the Homomorphism Interpolation Theorem in [HHP1] depends quite strongly on the bounds given in (12.16).

Theorem 12.30 For any graph G and any integer n between χ and ψ, there is a complete homomorphism (and hence a complete coloring) of G of order n.

Proof. Let $\psi(G) = t$ and let ϕ be a homomorphism of G onto K_t. If ϕ is just an isomorphism, then G is K_t and $\chi(G) = \psi(G)$. Otherwise, we can write $\phi = \varepsilon_m \cdots \varepsilon_2 \varepsilon_1$ where each ε_i is an elementary homomorphism. Let $G_1 = \varepsilon_1 G$, $G_2 = \varepsilon_2 G_1, \cdots, K_t = G_m = \varepsilon_m G_{m-1}$. We know from (12.16) that $\chi(G_{i+1}) \leq \chi(G_i) + 1$ for each i. Since $\chi(G_m) = \psi(G)$, it follows that for each n with $\chi(G) \leq n \leq t = \psi(G)$, there exists one graph in the sequence (G_i), say G_s, with chromatic number n. But then G_s has a complete homomorphism ϕ' of order n, and so $\phi' \varepsilon_s \cdots \varepsilon_2 \varepsilon_1$ is a homomorphism of G onto K_n.

Many upper bounds for $\chi(G)$ are also bounds for $\psi(G)$. As an example, we extend the upper bounds in (12.3) and (12.4), as in [HH1].

Theorem 12.31 For any graph G,

$$\psi + \bar{\chi} \le p + 1. \tag{12.18}$$

The next result follows from (12.18) and the fact that $\bar{\chi} \ge \beta_0$.

Corollary 12.31(a) For any graph G,

$$\psi \le p - \beta_0 + 1. \tag{12.19}$$

This inequality can also be proved directly using the proof of (12.3), which it sharpens.

THE CHROMATIC POLYNOMIAL

The chromatic polynomial of a graph was introduced by Birkhoff and Lewis [BL1] in their attack on the 4CC. Let G be a labeled graph. A *coloring of G from t colors* is a coloring of G which uses t or fewer colors. Two colorings of G from t colors will be considered different if at least one of the labeled points is assigned different colors.

Let us denote by $f(G, t)$ the number of different colorings of a labeled graph G from t colors. Of course $f(G, t) = 0$ if $t < \chi(G)$. Indeed the smallest t for which $f(G, t) > 0$ is the chromatic number of G. The 4CC therefore asserts that for every planar graph G, $f(G, 4) > 0$.

For example, there are t ways of coloring any given point of K_3. For a second point, any of $t - 1$ colors may be used, while there are $t - 2$ ways of coloring the remaining point. Thus

$$f(K_3, t) = t(t - 1)(t - 2).$$

This can be generalized to any complete graph,[*]

$$f(K_p, t) = t(t - 1)(t - 2) \cdots (t - p + 1) = t_{(p)}. \tag{12.20}$$

The corresponding polynomial of the totally disconnected graph $\bar{K}_p$ is particularly easy to find since each of its p points may be colored independently in any of t ways:

$$f(\bar{K}_p, t) = t^p. \tag{12.21}$$

The central point v_0 of $K_{1,4}$ in Fig. 12.14 may be colored in any of t ways while each endpoint may be colored in any of $t - 1$ ways. Therefore $f(K_{1,4}, t) = t(t - 1)^4$. In each of these examples, $f(G, t)$ is a polynomial in t. This is always the case, as we are about to see.

Theorem 12.32 If u and v are nonadjacent points in a graph G, and ε is the elementary homomorphism which identifies them, then

$$f(G, t) = f(G + uv, t) + f(\varepsilon G, t). \tag{12.22}$$

---

[*] Following Riordan [R15], we denote the expression for the falling factorial by $t_{(p)}$.

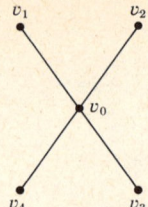

Fig. 12.14. A labeled copy of $K_{1,4}$.

Proof. The equation follows directly from two observations. First, the number of ways of coloring G from t colors where u and v are colored differently is precisely the number of ways of coloring $G + uv$ from t colors. Second, the number of ways of coloring G from t colors where u and v are colored the same is exactly the number of ways of coloring the homomorphic image εG from t colors, where ε identifies u and v.

This theorem now implies that if G is any noncomplete (p, q) graph, then there are graphs G_1 with $q + 1$ lines and G_2 with $p - 1$ points such that $f(G, t) = f(G_1, t) + f(G_2, t)$. The equation (12.22) can then be applied to G_1 and G_2, and so on, until only complete graphs are present. Hence $f(G, t)$ is the sum of expressions of the form $f(K_p, t)$. However $f(K_p, t) = t_{(p)}$ is a polynomial in t.

Corollary 12.32(a) For any graph G, $f(G, t)$ is a polynomial in t.

We thus refer to $f(G, t)$ as the *chromatic polynomial* of G. To illustrate the theorem, we employ a device introduced by Zykov [Z1] where a diagram of the graph is used to denote its chromatic polynomial, with t understood. We indicate by u and v the nonadjacent points considered at each step, following the exposition of Read [R6].

Thus for the graph G of Fig. 12.15,

$$f(G, t) = t_{(5)} + 3t_{(4)} + t_{(3)} = t^5 - 7t^4 + 18t^3 - 20t^2 + 8t.$$

In particular, the number of ways of coloring G from 3 colors is $f(G, 3) = 6$.

There are several properties of chromatic polynomials which now follow directly from Theorem 12.32.

Theorem 12.33 Let G be a graph with p points, q lines, and k components $G_1, G_2, \cdots, G_k$. Then

1. $f(G, t)$ has degree p.
2. The coefficient of t^p in $f(G, t)$ is 1.
3. The coefficient of t^{p-1} in $f(G, t)$ is $-q$.
4. The constant term in $f(G, t)$ is 0.
5. $f(G, t) = \Pi_{i=1}^{k} f(G_i, t)$.
6. The smallest exponent of t in $f(G, t)$ with a nonzero coefficient is k.

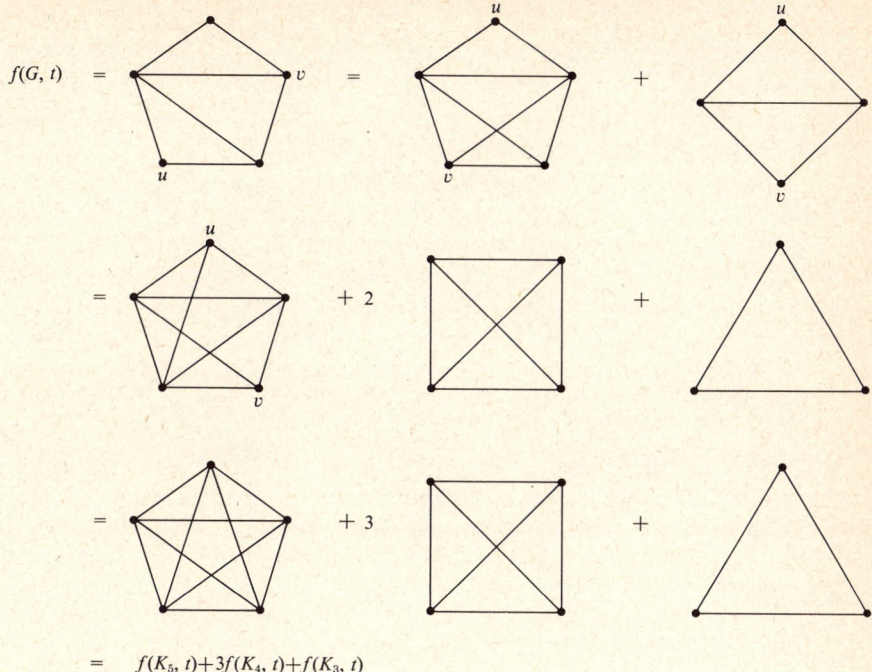

$$= \quad f(K_5, t) + 3f(K_4, t) + f(K_3, t)$$

Fig. 12.15. The determination of a chromatic polynomial.

Not quite so obvious is the following result discovered by Whitney [W10] and generalized by Rota [R20] using his powerful methods involving Möbius inversion.

Theorem 12.34 The coefficients of every chromatic polynomial alternate in sign.

Certainly, every two isomorphic graphs have the same chromatic polynomial. However, there are often several nonisomorphic graphs with the same chromatic polynomial; in fact, all trees with p points have equal chromatic polynomials.

Theorem 12.35 A graph G with p points is a tree if and only if

$$f(G, t) = t(t - 1)^{p-1}.$$

Proof. First we show that every labeled tree T with p points has $t(t - 1)^{p-1}$ as its chromatic polynomial. We proceed by induction on p, the result being obvious for $p = 1$ and $p = 2$. Assume the chromatic polynomial of all trees with $p - 1$ points is given by $t(t - 1)^{p-2}$. Let v be an endpoint of T and suppose $x = uv$ is the line of T incident with v. By hypothesis, the tree $T' = T - v$ has $t(t - 1)^{p-2}$ for its chromatic polynomial. The point v can be assigned any color different from that assigned to u, so that v may be colored in any of $t - 1$ ways. Thus $f(T, t) = (t - 1)f(T', t) = t(t - 1)^{p-1}$.

Conversely, let G be a graph such that $f(G, t) = t(t - 1)^{p-1}$. Since the coefficient of t in $f(G, t)$ is one, G is connected by Theorem 12.33(6). Furthermore, the coefficient of t^{p-1} is $-(p - 1)$ so that G has $p - 1$ lines by Theorem 12.33(3). Theorem 4.1 now guarantees that G is a tree.

It remains an unsolved problem to characterize graphs which have the same chromatic polynomial. Of a more basic nature is the unsolved problem of determining what polynomials are chromatic. For example, the polynomial $t^4 - 3t^3 + 3t^2$ satisfies all the known properties of a chromatic polynomial, but is not chromatic. For if it were $f(G, t)$ for some graph G, then necessarily G would have 4 points, 3 lines, and 2 components so that $G = K_3 \cup K_1$. However, the chromatic polynomial of this graph is

$$f(G, t) = t_{(3)}t = t^4 - 3t^3 + 2t^2.$$

It has been conjectured by Read [R6] that the absolute value of the coefficients of every chromatic polynomial are strictly increasing at first, then become strictly decreasing and remain so.

EXERCISES

12.1 Concerning the join of two graphs,

    a) $\chi(G_1 + G_2) = \chi(G_1) + \chi(G_2)$,
    b) G_1 and G_2 are critical if and only if their join $G_1 + G_2$ is.

12.2 If $n \geq 3$ is the length of the longest odd cycle of G, then $\chi(G) \leq n + 1$.

(Erdös and Hajnal [EH1])

12.3 If the points of G are labeled $v_1, v_2, \cdots, v_p$ so that $d_1 \geq d_2 \geq \cdots \geq d_p$, then $\chi(G) \leq \max_i \min \{i, d_i + 1\}$.   (Welsh and Powell [WP1])

12.4 If not every line lies on a hamiltonian cycle, then $\chi \leq 1 + p/2$.

12.5 The chromatic number of the conjunction $G_1 \wedge G_2$ of two graphs does not exceed that of either graph.   (S. T. Hedetniemi)

12.6 The only connected regular graph of degree $n \geq 3$ which is $(n + 1)$-chromatic is K_{n+1}.

12.7 The following regular graphs are all those for which the upper bounds in (12.4) and (12.5) are realized:

    a) $\chi + \bar{\chi} = p + 1$ only for K_p, $\bar{K}_p$, and C_5.
    b) $\chi\bar{\chi} = [((p + 1)/2)^2]$ only for K_1, $\bar{K}_2$, K_2, and C_5.   (Finck [F4])

12.8 a) If $p = p(G)$ is a prime, then $\chi\bar{\chi} = p$ only for $\bar{K}_p$ and K_p.
    b) $\chi^2 + \bar{\chi}^2 = p^2 + 1$ if and only if $G = K_p$ or $\bar{K}_p$; otherwise

$$\chi^2 + \bar{\chi}^2 \leq (p - 1)^2 + 4.$$

(Finck [F4])

12.9 Every outerplanar map is 3-colorable.

12.10 Every 4-connected plane map is 4-colorable.

12.11 In any coloring of a line-graph, each point is adjacent with at most two points of the same color.

12.12 Consider a connected graph G which is not an odd cycle. If all cycles have the same parity, then $\chi'(G) = \Delta(G)$. (J. A. Bondy and D. J. A. Welsh)

12.13 Find the line-chromatic numbers of K_p and of $K_{m,n}$.

(Behzad, Chartrand, and Cooper [BCC1])

12.14 If H is the graph obtained from G by taking $V(H) = X(G)$ and x, y are adjacent in H whenever they do not both lie in a complete subgraph of G, then $\chi(H)$ is the minimum number of complete subgraphs whose union is $V \cup X$. (Havel [H37])

12.15 Every toroidal graph has $\delta \le 6$, and hence has $\chi \le 7$.

12.16 There is a 5-critical graph with 9 points.

12.17 What is the smallest uniquely 3-colorable graph which is not complete?

12.18 What is the minimum number of lines in a uniquely n-colorable graph with p points? (Cartwright and Harary [CH2])

12.19 Obviously the chromatic number of any graph is at least as large as $\bar{\beta}_0$. For any odd cycle C_{2n+1}, $n \ge 2$, $\bar{\beta}_0$ is 2 and χ is 3. Construct a graph with no triangles, $\bar{\beta}_0 = 2$, and $\chi = 4$.
(This can be done with only 11 points.)

12.20 If $\chi(G) = n \ge 5$, then there are n points such that each pair are connected by at least four disjoint paths. (Dirac [D8])

12.21 For any integers d and n such that $1 < d \le n$, there exists an n-critical graph with $\bar{\beta}_0 = d$. (House [H47])

12.22 a) Every 3-chromatic maximal planar graph is uniquely 3-colorable.
  b) An outerplanar graph G with at least 3 points is uniquely 3-colorable if and only if it is maximal outerplanar. (Chartrand and Geller [CG1])

12.23 An n-critical graph cannot be separated by the points of a uniquely $(n-1)$-colorable subgraph. (Harary, Hedetniemi, and Robinson [HHR1])

12.24 For any independent set S of points of a critical graph G, $\chi(G - S) = \chi(G) - 1$.

(Dirac [D11])

12.25 For any elementary contraction η of a graph G, $|\chi(G) - \chi(\eta G)| \le 1$.

(Harary, Hedetniemi, and Prins [HHP1])

12.26 Determine the achromatic number of P_n, C_n, W_n, and $K_{m,n}$.

12.27 The n-chromatic number $\chi_n(G)$ is the smallest number m of colors needed to color G such that not all points on any path of length n are colored the same.

  a) For any n there is an outerplanar graph G such that $\chi_n(G) = 3$.
  b) For any n there is a planar graph G such that $\chi_n(G) = 4$.

(Chartrand, Geller, and Hedetniemi [CGH1])

12.28 If e is the length of a longest path in G then $\chi(G) \le e + 1$. (Gallai [G4])

12.29 The chromatic number of any graph G satisfies the lower bound

$$\chi(G) \ge p^2/(p^2 - 2q).$$

CHAPTER 13

MATRICES

In orderly disorder they
Wait coldly columned, dead, prosaic.
Poet, breathe on them and pray
They burn with life in your mosaic.

J. LUZZATO

A graph is completely determined by either its adjacencies or its incidences. This information can be conveniently stated in matrix form. Indeed, with a given graph, adequately labeled, there are associated several matrices, including the adjacency matrix, incidence matrix, cycle matrix, and cocycle matrix. It is often possible to make use of these matrices in order to identify certain properties of a graph. The classic theorem on graphs and matrices is the Matrix-Tree Theorem, which gives the number of spanning trees in any labeled graph. The matroids associated with the cycle and cocycle matrices of a graph are discussed.

THE ADJACENCY MATRIX

The *adjacency matrix* $A = [a_{ij}]$ *of a labeled graph* G with p points is the $p \times p$ matrix in which $a_{ij} = 1$ if v_i is adjacent with v_j and $a_{ij} = 0$ otherwise. Thus there is a one-to-one correspondence between labeled graphs with p points and $p \times p$ symmetric binary matrices with zero diagonal.

Figure 13.1 shows a labeled graph G and its adjacency matrix A. One immediate observation is that the row sums of A are the degrees of the points of G. In general, because of the correspondence between graphs and matrices, any graph-theoretic concept is reflected in the adjacency matrix. For example, recall from Chapter 2 that a graph G is connected if and only if there is no partition $V = V_1 \cup V_2$ of the points of G such that no line joins a point of V_1 with a point of V_2. In matrix terms we may say that G is connected if and only if there is no labeling of the points of G such that its adjacency matrix has the reduced form

$$A = \begin{bmatrix} A_{11} & 0 \\ 0 & A_{22} \end{bmatrix},$$

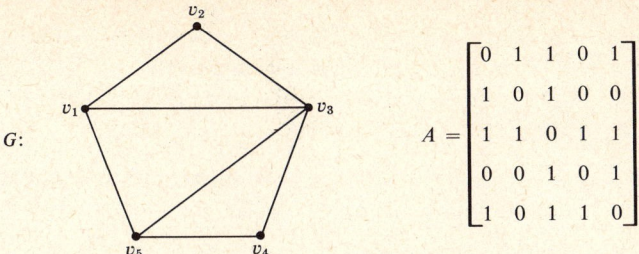

$$A = \begin{bmatrix} 0 & 1 & 1 & 0 & 1 \\ 1 & 0 & 1 & 0 & 0 \\ 1 & 1 & 0 & 1 & 1 \\ 0 & 0 & 1 & 0 & 1 \\ 1 & 0 & 1 & 1 & 0 \end{bmatrix}$$

Fig. 13.1. A labeled graph and its adjacency matrix.

where A_{11} and A_{22} are square. If A_1 and A_2 are adjacency matrices which correspond to two different labelings of the same graph G, then for some permutation matrix P, $A_1 = P^{-1}A_2P$. Sometimes a labeling is irrelevant, as in the following results which interpret the entries of the powers of the adjacency matrix.

Theorem 13.1 Let G be a labeled graph with adjacency matrix A. Then the i, j entry of A^n is the number of walks of length n from v_i to v_j.

Corollary 13.1(a) For $i \neq j$, the i, j entry of A^2 is the number of paths of length 2 from v_i to v_j. The i, i entry of A^2 is the degree of v_i and that of A^3 is twice the number of triangles containing v_i.

Corollary 13.1(b) If G is connected, the distance between v_i and v_j for $i \neq j$ is the least integer n for which the i, j entry of A^n is nonzero.

The *adjacency matrix of a labeled digraph D* is defined similarly: $A = A(D) = [a_{ij}]$ has $a_{ij} = 1$ if arc v_iv_j is in D and is 0 otherwise. Thus $A(D)$ is not necessarily symmetric. Some results for digraphs using $A(D)$ will be given in Chapter 16. By definition of $A(D)$, the adjacency matrix of a given graph can also be regarded as that of a symmetric digraph. We now apply this observation to investigate the determinant of the adjacency matrix of a graph, following [H27].

A *linear subgraph of a digraph D* is a spanning subgraph in which each point has indegree one and outdegree one. Thus it consists of a disjoint spanning collection of directed cycles.

Theorem 13.2 If D is a digraph whose linear subgraphs are D_i, $i = 1, \cdots, n$, and D_i has e_i even cycles, then

$$\det A(D) = \sum_{i=1}^{n} (-1)^{e_i}.$$

Every graph G is associated with that digraph D with arcs v_iv_j and v_jv_i whenever v_i and v_j are adjacent in G. Under this correspondence, each linear subgraph of D yields a spanning subgraph of G consisting of a point disjoint collection of lines and cycles, which is called a *linear subgraph of a graph*.

Those components of a linear subgraph of G which are lines correspond to the 2-cycles in the linear subgraph of D in a one-to-one fashion, but those components which are cycles of G correspond to two directed cycles in D. Since $A(G) = A(D)$ when G and D are related as above, the determinant of $A(G)$ can be calculated.

Corollary 13.2(a) If G is a graph whose linear subgraphs are G_i, $i = 1, \cdots, n$, where G_i has e_i even components and c_i cycles, then

$$\det A(G) = \sum_{i=1}^{n} (-1)^{e_i} 2^{c_i}.$$

THE INCIDENCE MATRIX

A second matrix, associated with a graph G in which the points and lines are labeled, is the *incidence matrix* $B = [b_{ij}]$. This $p \times q$ matrix has $b_{ij} = 1$ if v_i and x_j are incident and $b_{ij} = 0$ otherwise. As with the adjacency matrix, the incidence matrix determines G up to isomorphism. In fact any $p - 1$ rows of B determine G since each row is the sum of all the others modulo 2.

The next theorem relates the adjacency matrix of the line graph of G to the incidence matrix of G. We denote by B^T the transpose of matrix B.

Theorem 13.3 For any (p, q) graph G with incidence matrix B,

$$A(L(G)) = B^T B - 2I_q.$$

Let M denote the matrix obtained from $-A$ by replacing the ith diagonal entry by deg v_i. The following theorem is contained in the pioneering work of Kirchhoff [K7].

Theorem 13.4 (Matrix-Tree Theorem) Let G be a connected labeled graph with adjacency matrix A. Then all cofactors of the matrix M are equal and their common value is the number of spanning trees of G.

Proof. We begin the proof by changing either of the two 1's in each column of the incidence matrix B of G to -1, thereby forming a new matrix E. (We will see in Chapter 16 that this amounts to arbitrarily orienting the lines of G and taking E as the incidence matrix of this oriented graph.)

The i, j entry of EE^T is $e_{i1}e_{j1} + e_{i2}e_{j2} + \cdots + e_{iq}e_{jq}$, which has the value deg v_i if $i = j$, -1 if v_i and v_j are adjacent, and 0 otherwise. Hence $EE^T = M$.

Consider any submatrix of E consisting of $p - 1$ of its columns. This $p \times (p - 1)$ matrix corresponds to a spanning subgraph H of G having $p - 1$ lines. Remove an arbitrary row, say the kth, from this matrix to obtain a square matrix F of order $p - 1$. We will show that $|\det F|$ is 1 or 0 according as H is or is not a tree. First, if H is not a tree, then because H has p points and $p - 1$ lines, it is disconnected, implying that there is a

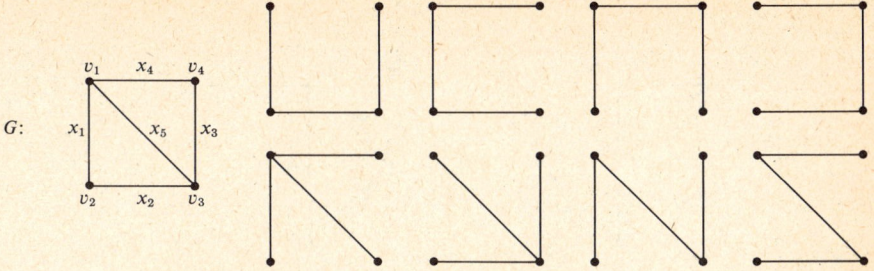

Fig. 13.2. $K_4 - x$ and its spanning trees.

component not containing v_k. Since the rows corresponding to the points of this component are dependent, det $F = 0$. On the other hand, suppose H is a tree. In this case, we can relabel its lines and points other than v_k as follows: Let $u_1 \neq v_k$ be an endpoint of H (whose existence is guaranteed by Corollary 4.1(a)), and let y_1 be the line incident with it; let $u_2 \neq v_k$ be any endpoint of $H - u_1$ and y_2 its incident line, and so on. This relabeling of the points and lines of H determines a new matrix F' which can be obtained by permuting the rows and columns of F independently. Thus $|\det F'| = |\det F|$. However, F' is lower triangular with every diagonal entry $+1$ or -1; hence, $|\det F| = 1$.

The following algebraic result, usually called the Binet-Cauchy Theorem, will now be very useful.

Lemma 13.4(a) If P and Q are $m \times n$ and $n \times m$ matrices, respectively, with $m \leq n$, then det PQ is the sum of the products of corresponding major determinants of P and Q.

(A major determinant of P or Q has order m, and the phrase "corresponding major determinants" means that the columns of P in the one determinant are numbered like the rows of Q in the other.)

We apply this lemma to calculate the first principal cofactor of M. Let E_1 be the $(p - 1) \times q$ submatrix obtained from E by striking out its first row. By letting $P = E_1$ and $Q = E_1^T$, we find, from the lemma, that the first principal cofactor of M is the sum of the products of the corresponding major determinants of E_1 and E_1^T. Obviously, the corresponding major determinants have the same value. We have seen that their product is 1 if the columns from E_1 correspond to a spanning tree of G and is 0 otherwise. Thus the sum of these products is exactly the number of spanning trees.

The equality of all the cofactors, both principal and otherwise, holds for every matrix whose row sums and column sums are all zero, completing the proof.

To illustrate the Matrix-Tree Theorem, we consider a labeled graph G taken at random, say $K_4 - x$. This graph, shown in Fig. 13.2, has eight

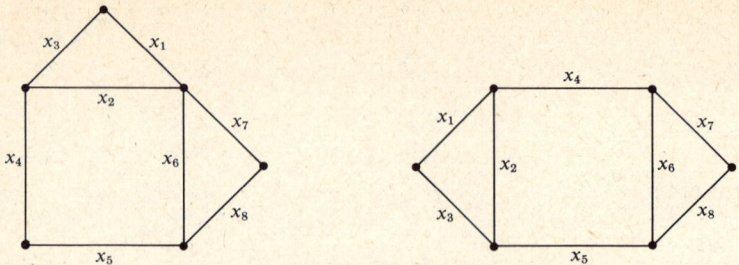

Fig. 13.3. Two graphs with the same cycle matrix.

spanning trees, since the 2,3 cofactor, for example,

$$\text{of}\quad M = \begin{bmatrix} 3 & -1 & -1 & -1 \\ -1 & 2 & -1 & 0 \\ -1 & -1 & 3 & -1 \\ -1 & 0 & -1 & 2 \end{bmatrix} \quad \text{is} \quad -\begin{vmatrix} 3 & -1 & -1 \\ -1 & -1 & -1 \\ -1 & 0 & 2 \end{vmatrix} = 8.$$

The number of labeled trees with p points is easily found by applying the Matrix-Tree Theorem to K_p. Each principal cofactor is the determinant of order $p - 1$:

$$\begin{vmatrix} p-1 & -1 & \cdots & -1 \\ -1 & p-1 & \cdots & -1 \\ \cdot & & \cdot & \\ \cdot & & \cdot & \\ \cdot & & \cdot & \\ -1 & -1 & \cdots & p-1 \end{vmatrix}$$

Subtracting the first row from each of the others and adding the last $p - 2$ columns to the first yields an upper triangular matrix whose determinant is p^{p-2}.

Corollary 13.4(a) The number of labeled trees with p points is p^{p-2}.

There appear to be as many different ways of proving this formula as there are independent discoveries thereof. An interesting compilation of such proofs is presented in Moon [M15].

THE CYCLE MATRIX

Let G be a graph whose lines and cycles are labeled. The *cycle matrix* $C = [c_{ij}]$ of G has a row for each cycle and a column for each line with $c_{ij} = 1$ if the ith cycle contains line x_j and $c_{ij} = 0$ otherwise. In contrast to the adjacency and incidence matrices, the cycle matrix does not determine a graph up to isomorphism. Obviously the presence or absence of lines which lie on no cycle is not indicated. Even when such lines are excluded, however, C does not determine G, as is shown by the pair of graphs in Fig. 13.3, which both have cycles

$$Z_1 = \{x_1, x_2, x_3\} \qquad\qquad Z_2 = \{x_2, x_4, x_5, x_6\}$$
$$Z_3 = \{x_6, x_7, x_8\} \qquad\qquad Z_4 = \{x_1, x_3, x_4, x_5, x_6\}$$
$$Z_5 = \{x_2, x_4, x_5, x_7, x_8\} \qquad Z_6 = \{x_1, x_3, x_4, x_5, x_7, x_8\}$$

and therefore share the cycle matrix

$$
C =
\begin{array}{c}
\begin{array}{cccccccc}
x_1 & x_2 & x_3 & x_4 & x_5 & x_6 & x_7 & x_8
\end{array} \\
\left[
\begin{array}{cccccccc}
1 & 1 & 1 & 0 & 0 & 0 & 0 & 0 \\
0 & 1 & 0 & 1 & 1 & 1 & 0 & 0 \\
0 & 0 & 0 & 0 & 0 & 1 & 1 & 1 \\
1 & 0 & 1 & 1 & 1 & 1 & 0 & 0 \\
0 & 1 & 0 & 1 & 1 & 0 & 1 & 1 \\
1 & 0 & 1 & 1 & 1 & 0 & 1 & 1
\end{array}
\right]
\begin{array}{c}
Z_1 \\ Z_2 \\ Z_3 \\ Z_4 \\ Z_5 \\ Z_6
\end{array}
\end{array}
$$

The next theorem provides a relationship between the cycle and incidence matrices. In combinatorial topology this result is described by saying that the boundary of the boundary of any chain is zero.

Theorem 13.5 If G has incidence matrix B and cycle matrix C, then

$$CB^T \equiv 0 \ (\text{mod } 2).$$

Proof. Consider the ith row of C and jth column of B^T, which is the jth row of B. The rth entries in these two rows are both nonzero if and only if x_r is in the ith cycle Z_i and is incident with v_j. If x_r is in Z_i, then v_j is also, but if v_j is in the cycle, then there are two lines of Z_i incident with v_j so that the i, j entry of CB^T is $1 + 1 \equiv 0 \ (\text{mod } 2)$.

Analogous to the cycle matrix, one can define the *cocycle matrix* $C^*(G)$. If G is 2-connected, then each point of G corresponds to the cocycle (minimal cutset) consisting of the lines incident with it. Therefore, the incidence matrix of a block is contained in its cocycle matrix.

Since every row of the incidence matrix B is the sum modulo 2 of the other rows, it is clear that the rank of B is at most $p - 1$. On the other hand, if the rank of B is less than $p - 1$, then there is some set of fewer than p rows whose sum, modulo 2, is zero. But then there can be no line joining a point in the set belonging to those rows and a point not in that set, so G cannot be connected. Thus we have one part of the next theorem. The other parts follow directly from the results in Chapter 4 which give the dimensions of the cycle and cocycle spaces of G.

Theorem 13.6 For a connected graph G, the ranks of the cycle, incidence, and cocycle matrices are $r(C) = q - p + 1$ and $r(B) = r(C^*) = p - 1$.

In view of Theorem 13.6, an important submatrix of the cycle matrix C of a connected graph is given by any $m = q - p + 1$ rows representing a

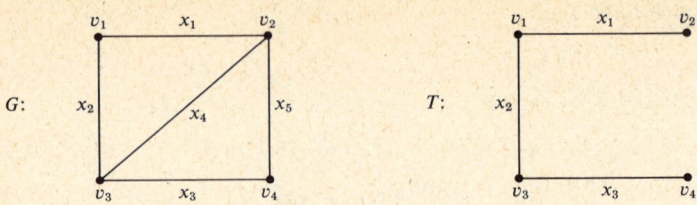

Fig. 13.4. A graph and a spanning tree.

cycle basis. Each such reduced matrix $C_0(G)$ is an $m \times q$ submatrix of C, and similarly a reduced cocyle matrix $C_0^*(G)$ is $m^* \times q$, where $m^* = p - 1$. Then by Theorem 13.5, we have immediately $CC^{*T} \equiv 0 \pmod 2$ and hence also $C_0 C_0^{*T} \equiv 0 \pmod 2$. A reduced incidence matrix B_0 is obtained from B by deletion of the last row. By an earlier remark, no information is lost by so reducing B.

If the cycles and cocycles are chosen in a special way, then the reduced incidence, cycle, and cocycle matrices of a graph have particularly nice forms. Recall from Chapter 4 that any spanning tree T determines a cycle basis and a cocycle basis for G. In particular, if $X_1 = \{x_1, x_2, \cdots, x_{p-1}\}$ is the set of twigs (lines) of T, and $X_2 = \{x_p, x_{p+1}, \cdots, x_q\}$ is the set of its chords, then there is a unique cycle Z_i in $G - X_2 + x_i$, $p \le i \le q$, and a unique cocycle Z_j^* in $G - X_1 + x_j$, $1 \le j \le p - 1$, and these collections of cycles and cocycles form bases for their respective spaces. For example, in the graph G of Fig. 13.4 the cycles and cocycles determined by the particular spanning tree T shown are

$$Z_4 = \{x_1, x_2, x_4\}, \qquad Z_1^* = \{x_1, x_4, x_5\},$$
$$Z_5 = \{x_1, x_2, x_3, x_5\}, \qquad Z_2^* = \{x_2, x_4, x_5\},$$
$$Z_3^* = \{x_3, x_5\}.$$

The reduced matrices, which are determined both by G and the choice of T, are:

$$
B_0(G, T) = \begin{array}{c} \\ v_1 \\ v_2 \\ v_3 \end{array}
\begin{array}{cc} X_1 & X_2 \\ \overbrace{x_1 \ x_2 \ x_3} \ \overbrace{x_4 \ x_5} \\ \left[\begin{array}{ccc|cc} 1 & 1 & 0 & 0 & 0 \\ 1 & 0 & 0 & 1 & 1 \\ 0 & 1 & 1 & 1 & 0 \end{array} \right] \end{array},
$$

$$
C_0(G, T) = \begin{array}{c} \\ Z_4 \\ Z_5 \end{array}
\begin{array}{cc} X_1 & X_2 \\ \overbrace{\quad\quad} \ \overbrace{\quad} \\ \left[\begin{array}{ccc|cc} 1 & 1 & 0 & 1 & 0 \\ 1 & 1 & 1 & 0 & 1 \end{array} \right] \end{array},
$$

and

$$C_0^*(G, T) = \begin{array}{c} Z_1^* \\ Z_2^* \\ Z_3^* \end{array} \overset{\displaystyle\overbrace{}^{X_1}\ \overbrace{}^{X_2}}{\begin{bmatrix} 1 & 0 & 0 & 1 & 1 \\ 0 & 1 & 0 & 1 & 1 \\ 0 & 0 & 1 & 0 & 1 \end{bmatrix}}.$$

It is easy to see that this is a special case of the following equations (all modulo 2) which hold for any connected graph G and spanning tree T:

$$B_0 = B_0(G, T) = \overset{X_1 \quad X_2}{[\widetilde{B_1} \quad \widetilde{B_2}]}, \qquad C_0 = C_0(G, T) = \overset{X_1 \quad X_2}{[\widetilde{C_1} \quad I_m]},$$

and

$$C_0^* = C_0^*(G, T) = \overset{X_1 \quad X_2}{[\widetilde{I_{m*}} \quad \widetilde{C_2^*}]},$$

where $C_1^T = B_1^{-1}B_2 = C_2^*$ and $C_0^* = B_1^{-1}B_0 = [I_{m*} \quad C_1^T]$. It follows from these equations that, given G and T, each of the partitioned matrices B_0, C_0, and C_0^* determines the other two.

Excursion—Matroids Revisited

The cycle and cocycle matrices are particular representations of the cycle matroid and cocycle matroid of a graph, introduced in Chapter 4. A matroid is called *graphical* if it is the cycle matroid of some graph, and *cographical* if it is a cocycle matroid. Tutte [T12] has determined which matroids are graphical or cographical, thereby inadvertently solving a previously open problem in electric network theory.

The smallest example of a matroid which is not graphical or cographical is the self-dual matroid obtained by taking $M = \{1, 2, 3, 4\}$ and the circuits all 3-element subsets of M.

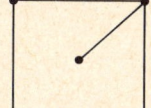 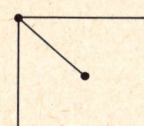 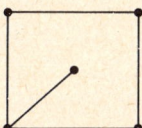 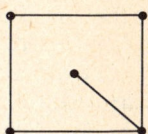

Fig. 13.5. The new circuits in the whirl of W_5.

Another example, Tutte [T19], of a matroid which is not graphical involves the wheel $W_{n+1} = K_1 + C_n$. Its cycle matroid has $n^2 - n + 1$ circuits since there are that many cycles in a wheel. If in this matroid we remove from the collection of circuits the cycle C_n which forms the rim of the wheel, and add to it all of the "spoked rims" (the sets of lines in the subgraphs shown in Fig. 13.5), then it can be shown that the result is a new matroid which is not graphical or cographical. This is called a *whirl* of order n and is generated by n^2 circuits.

Even if a matroid is graphical, it need not be cographical. For example, the cycle matroid of K_5 is not cographical. In fact a matroid is both graphical and cographical if and only if it is the cycle matroid of some planar graph.

EXERCISES

13.1 a) Characterize the adjacency matrix of a bipartite graph.
   b) A graph G is bipartite if and only if for all odd n every diagonal entry of A^n is 0.

13.2 Let G be a connected graph with adjacency matrix A. What can be said about A if

   a) v_i is a cutpoint?
   b) v_iv_j is a bridge?

13.3 If $c_n(G)$ is the number of n-cycles of a graph G with adjacency matrix A, then

   a) $c_3(G) = \frac{1}{6}\mathrm{tr}(A^3)$.
   b) $c_4(G) = \frac{1}{8}[\mathrm{tr}(A^4) - 2q - 2 \Sigma_{i \neq j} a_{ij}^{(2)}]$.
   c) $c_5(G) = \frac{1}{10}[\mathrm{tr}(A^5) - 5\,\mathrm{tr}(A^3) - 5 \Sigma_{i=1}^{p}(\Sigma_{j=1}^{p} a_{ij} - 2)a_{ii}^{(3)}]$.

(Harary and Manvel [HM1])

13.4 a) If G is a disconnected labeled graph, then every cofactor of M is 0.
   b) If G is connected, the number of spanning trees of G is the product of the number of spanning trees of the blocks of G.

(Brooks, Smith, Stone, and Tutte [BSST1])

13.5 Let G be a labeled graph with lines $x_1, x_2, \cdots, x_q$. Define the $p \times p$ matrix $M_x = [m_{ij}]$ by

$$m_{ij} = \begin{cases} -x_k & \text{if} \quad x_k = v_iv_j \\ 0 & \text{if} \quad v_i \text{ and } v_j \text{ are not adjacent} \end{cases} \quad \text{for} \quad i \neq j,$$

$$-m_{ii} = \sum_{n \neq i} m_{in}.$$

By the term of a spanning tree of G is meant the product of its lines. The tree polynomial of G is defined as the sum of the terms of its spanning trees.

The Variable Matrix Tree Theorem asserts that the value of any cofactor of the matrix M_x is the tree polynomial of G.

13.6 Do there exist two different graphs with the same cycle matrix which are smaller than those in Fig. 13.3?

13.7 The "cycle-matroid" and "cocycle-matroid" of a graph do indeed satisfy the first definition of matroid given in Chapter 4.

13.8 Two graphs G_1 and G_2 are *cospectral* if the polynomials $\det(A_1 - tI)$ and $\det(A_2 - tI)$ are equal. There are just two different cospectral graphs with 5 points.

(F. Harary, C. King, and R. C. Read)

13.9 If the eigenvalues of $A(G)$ are distinct, then every nonidentity automorphism of G has order 2.

(Mowshowitz [M17])

13.10 Let $f(t)$ be a polynomial of minimum degree (if any) such that every entry of $f(A)$ is 1, where A is the adjacency matrix of G. Then a graph has such a polynomial if and only if it is connected and regular.   (Hoffman [H45])

13.11 An *eulerian matroid* has a partition of its set S of elements into circuits.

  a) A graphical matroid is eulerian if and only if it is the cycle matroid of an eulerian graph,
  b) Not every eulerian matroid is graphical.

13.12 In a *binary* matroid, the intersection of every circuit and cocircuit has even cardinality. Every cocircuit of a binary eulerian matroid has even cardinality. In other words, the dual of a binary eulerian matroid is a "bipartite matroid," defined as expected.

(Welsh [W9])

CHAPTER 14

GROUPS

Tyger! Tyger! burning bright
In the forests of the night,
What immortal hand or eye
Could frame thy fearful symmetry?
WILLIAM BLAKE

From its inception, the theory of groups has provided an interesting and powerful abstract approach to the study of the symmetries of various configurations. It is not surprising that there is a particularly fruitful interaction between groups and graphs. In order to place the topic in its proper setting, we recall some elementary but relevant facts about groups. In particular, we develop several operations on permutation groups. These operations play an important role in graph theory as they are closely related to operations on graphs and are fundamental in graphical enumeration.

Any model of a given axiom system has an automorphism group, and graphs are no exception. It is observed that the group of a composite graph may be characterized in terms of the groups of its constituent graphs under suitable circumstances. Results are also presented on the existence of a graph with given group and given structural properties. The chapter is concluded with a study of graphs which are symmetric with respect to their points or lines.

THE AUTOMORPHISM GROUP OF A GRAPH

First we recall the usual definition of a group. The nonempty set A together with a binary operation, denoted by the juxtaposition $\alpha_1\alpha_2$ for α_1, α_2 in A, constitutes a *group* whenever the following four axioms are satisfied:

Axiom 1 (closure) For all α_1, α_2 in A, $\alpha_1\alpha_2$ is also an element of A.

Axiom 2 (associativity) For all α_1, α_2, α_3 in A,

$$\alpha_1(\alpha_2\alpha_3) = (\alpha_1\alpha_2)\alpha_3.$$

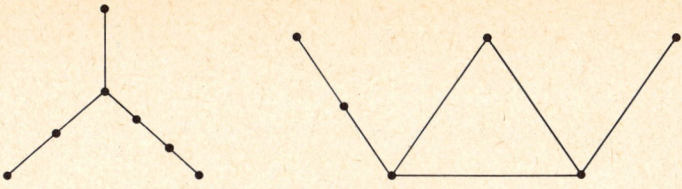

Fig. 14.1. Two identity graphs.

Axiom 3 (identity) There is an element i in A such that

$$i\alpha = \alpha i = \alpha \qquad \text{for all } \alpha \text{ in } A.$$

Axiom 4 (inversion) If Axiom 3 holds, then for each α in A, there is an element denoted α^{-1} such that

$$\alpha\alpha^{-1} = \alpha^{-1}\alpha = i.$$

A 1–1 mapping from a finite set onto itself is called a *permutation*. The usual composition of mappings provides a binary operation for permutations on the same set. Furthermore, whenever a collection of permutations is closed with respect to this composition, Axioms 2, 3, and 4 are automatically satisfied and it is called a *permutation group*. If a permutation group A acts on object set X, then $|A|$ is the *order* of this group and $|X|$ is the *degree*.

When A and B are permutation groups acting on the sets X and Y respectively, we will write $A \cong B$ to mean that A and B are *isomorphic groups*. However $A \equiv B$ indicates not only isomorphism but that A and B are *identical permutation groups*. More specifically $A \cong B$ if there is a 1–1 map $h: A \leftrightarrow B$ between the permutations such that for all α_1, α_2 in A, $h(\alpha_1\alpha_2) = h(\alpha_1)h(\alpha_2)$. To define $A \equiv B$ precisely, we also require another 1–1 map $f: X \leftrightarrow Y$ between the objects such that for all x in X and α in A, $f(\alpha x) = h(\alpha)f(x)$.

An *automorphism* of a graph G is an isomorphism of G with itself. Thus each automorphism α of G is a permutation of the point set V which preserves adjacency. Of course, α sends any point onto another of the same degree. Obviously any automorphism followed by another is also an automorphism, hence the automorphisms of G form a permutation group, $\Gamma(G)$, which acts on the points of G. It is known as *the group* of G, or sometimes as the *point-group* of G. The group $\Gamma(D)$ of a digraph D is defined similarly.

The identity map from V onto V is of course always an automorphism of G. For some graphs, it is the only automorphism; these are called *identity graphs*. The smallest nontrivial identity tree has seven points and is shown in Fig. 14.1, as is an identity graph with six points.

The point-group of G induces another permutation group $\Gamma_1(G)$, called the *line-group* of G, which acts on the lines of G. To illustrate the difference between these two groups, consider $K_4 - x$ shown in Fig. 14.2 with points

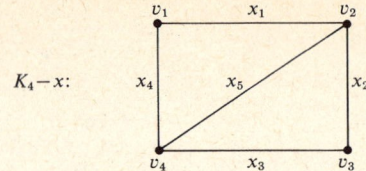

Fig. 14.2. A graph with labeled points and
lines.

$K_4 - x$:

labeled v_1, v_2, v_3, v_4 and lines x_1, x_2, x_3, x_4, x_5. The point-group $\Gamma(K_4 - x)$ consists of the four permutations:

$$(v_1)(v_2)(v_3)(v_4), \qquad (v_1)(v_3)(v_2v_4), \qquad (v_2)(v_4)(v_1v_3), \qquad (v_1v_3)(v_2v_4).$$

The identity permutation of the point-group induces the identity permutation on the lines, while $(v_1)(v_3)(v_2v_4)$ induces a permutation on the lines which fixes x_5, interchanges x_1 with x_4 and x_2 with x_3. In this way, one sees that the line-group $\Gamma_1(K_4 - x)$ consists of the following permutations, induced respectively by the above members of the point-group:

$$(x_1)(x_2)(x_3)(x_4)(x_5), \quad (x_1x_4)(x_2x_3)(x_5), \quad (x_1x_2)(x_3x_4)(x_5), \quad (x_1x_3)(x_2x_4)(x_5).$$

Of course the line-group and the point-group of $K_4 - x$ are isomorphic. But they are certainly not identical permutation groups since $\Gamma_1(K_4 - x)$ has degree 5 and $\Gamma(K_4 - x)$ has degree 4. Note that the line x_5 is fixed by every member of the line-group. Even the permutation group obtained from $\Gamma_1(K_4 - x)$ by restricting its object set to x_1, x_2, x_3, x_4 is not identical with $\Gamma(K_4 - x)$, since these two isomorphic permutation groups of the same degree have different cycle structure. Furthermore, it can be shown that even when two permutation groups have the same degree and the same cycle structure, they still need not be identical; see Pólya [P5, p. 176].

The next theorem [HP15] answers the question: when are $\Gamma(G)$ and $\Gamma_1(G)$ isomorphic? Sabidussi [S1] demonstrated the sufficiency using group theoretic methods.

Theorem 14.1 The line-group and the point-group of a graph G are isomorphic if and only if G has at most one isolated point and K_2 is not a component of G.

Proof. Let α' be the permutation in $\Gamma_1(G)$ which is induced by the permutation α in $\Gamma(G)$. By the definition of multiplication in $\Gamma_1(G)$, we have

$$\alpha'\beta' = (\alpha\beta)'$$

for all α, β in $\Gamma(G)$. Thus the mapping $\alpha \to \alpha'$ is a group homomorphism from $\Gamma(G)$ onto $\Gamma_1(G)$. Hence $\Gamma(G) \cong \Gamma_1(G)$ if and only if the kernel of this mapping is trivial.

To prove the necessity, assume $\Gamma(G) \cong \Gamma_1(G)$. Then $\alpha \neq i$ (the identity permutation) implies $\alpha' \neq i$. If G has distinct isolated points v_1 and v_2, we can define $\alpha \in \Gamma(G)$ by $\alpha(v_1) = v_2$, $\alpha(v_2) = v_1$, and $\alpha(v) = v$ for all $v \neq v_1, v_2$.

Then $\alpha \neq i$ but $\alpha' = i$. If K_2 is a component of G, take the line of K_2 to be $x = v_1 v_2$ and define $\alpha \in \Gamma(G)$ exactly as above to obtain $\alpha \neq i$ but $\alpha' = i$.

To prove the sufficiency, assume that G has at most one isolated point and that K_2 is not a component of G. If $\Gamma(G)$ is trivial, then obviously $\Gamma_1(G)$ fixes every line and hence $\Gamma_1(G)$ is trivial. Therefore, suppose there exists $\alpha \in \Gamma(G)$ with $\alpha(u) = v \neq u$. Then the degree of u is equal to the degree of v. Since u and v are not isolated, this degree is not zero.

CASE 1. u is adjacent to v. Let $x = uv$. Since K_2 is not a component, the degrees of both u and v are greater than one. Hence there is a line $y \neq x$ which is incident with u and $\alpha'(y)$ is incident with v. Therefore $\alpha'(y) \neq y$ and so $\alpha' \neq i$.

CASE 2. u is not adjacent to v. Let x be any line incident with u. Then $\alpha'(x) \neq x$ and so $\alpha' \neq i$, completing the proof.

OPERATIONS ON PERMUTATION GROUPS

There are several important operations on permutation groups which produce other permutation groups. We now develop four such binary operations: sum, product, composition, and power group.

Let A be a permutation group of order $m = |A|$ and degree d acting on the set $X = \{x_1, x_2, \cdots, x_d\}$, and let B be another permutation group of order $n = |B|$ and degree e acting on the set $Y = \{y_1, y_2, \cdots, y_e\}$. For example, let $A = C_3$, the cyclic group of degree 3, which acts on $X = \{1, 2, 3\}$. Then the three permutations of C_3 may be written $(1)(2)(3)$, (123), and (132). With $B = S_2$, the symmetric group of degree 2, acting on $Y = \{a, b\}$, we have the permutations $(a)(b)$ and (ab). We will use these two permutation groups to illustrate the binary operations defined here.

Their *sum** $A + B$ is a permutation group which acts on the disjoint union $X \cup Y$ and whose elements are all the ordered pairs of permutations α in A and β in B, written $\alpha + \beta$. Any object z of $X \cup Y$ is permuted by $\alpha + \beta$ according to the rule:

$$(\alpha + \beta)(z) = \begin{cases} \alpha z, & z \in X \\ \beta z, & z \in Y \end{cases}. \tag{14.1}$$

Thus $C_3 + S_2$ contains 6 permutations each of which can be written as the sum of permutations $\alpha \in C_3$ and $\beta \in S_2$ such as $(123)(ab) = (123) + (ab)$.

The *product*** $A \times B$ of A and B is a permutation group which acts on the set $X \times Y$ and whose permutations are all the ordered pairs, written $\alpha \times \beta$, of permutations α in A and β in B. The object (x, y) of $X \times Y$ is

---

* Sometimes called product or direct product and denoted accordingly.

** Also known as cartesian product; see [H18].

Table 14.1

OPERATIONS ON PERMUTATION GROUPS

			Sum	Product	Composition	Power
group	A	B	$A + B$	$A \times B$	$A[B]$	B^A
objects	X	Y	$X \cup Y$	$X \times Y$	$X \times Y$	Y^X
order	m	n	mn	mn	mn^d	mn
degree	d	e	$d + e$	de	de	e^d

permuted by $\alpha \times \beta$ as expected:

$$(\alpha \times \beta)(x, y) = (\alpha x, \beta y). \tag{14.2}$$

The product $C_3 \times S_2$ also has order 6 but while the degree of the sum $C_3 + S_2$ is 5, that of the product is 6. The permutation in $C_3 \times S_2$ corresponding to $(123) + (ab)$ in the sum is $(1a\ 2b\ 3a\ 1b\ 2a\ 3b)$, where for brevity $1a$ denotes $(1, a)$.

The *composition** $A[B]$ of "A around B" also acts on $X \times Y$. For each α in A and any sequence $(\beta_1, \beta_2, \cdots, \beta_d)$ of d (not necessarily distinct) permutations in B, there is a unique permutation in $A[B]$ written $(\alpha; \beta_1, \beta_2, \cdots, \beta_d)$ such that for (x_i, y_j) in $X \times Y$:

$$(\alpha; \beta_1, \beta_2, \cdots, \beta_d)(x_i, y_j) = (\alpha x_i, \beta_i y_j). \tag{14.3}$$

The composition $C_3[S_2]$ has degree 6 but its order is 24. Each permutation in $C_3[S_2]$ may be written in the form in which it acts on $X \times Y$. Using the same notation $1a$ for the ordered pair $(1, a)$ and applying the definition (14.3), one can verify that $((123); (a)(b), (ab), (a)(b))$ is expressible as $(1a\ 2a\ 3b\ 1b\ 2b\ 3a)$. Note that $S_2[C_3]$ has order 18 and so is not isomorphic to $C_3[S_2]$.

The *power group*** denoted by B^A acts on Y^X, the set of all functions from X into Y. We will always assume that the power group acts on more than one function. For each pair of permutations α in A and β in B there is a unique permutation, written β^α in B^A. We specify the action of β^α on any function f in Y^X by the following equation which gives the image of each $x \in X$ under the function $\beta^\alpha f$:

$$(\beta^\alpha f)(x) = \beta f(\alpha x). \tag{14.4}$$

The power group $S_2^{C_3}$ has order 6 and degree 8. It is easy to see by applying (14.4) that the permutation in this group obtained from $\alpha = (123)$ and $\beta = (ab)$ has one cycle of length 2 and one of length 6.

Table 14.1 summarizes the information concerning the order and degree of each of these four operations.

---

* Called "Gruppenkranz" by Pólya[P6] and "wreath product" by Littlewood[L3] and others.
** Not called by any other name as yet.

Table 14.2

PERMUTATION GROUPS OF DEGREE p

	Symbol	Order	Definition
Symmetric	S_p	$p!$	All permutations on $\{1, 2, \cdots, p\}$
Alternating	A_p	$p!/2$	All even permutations on $\{1, 2, \cdots, p\}$
Cyclic	C_p	p	Generated by $(12 \cdots p)$
Dihedral	D_p	$2p$	Generated by $(12 \cdots p)$ and $(1p)(2\,p-1) \cdots$
Identity	E_p	1	$(1)(2) \cdots (p)$ is the only permutation.

We now see that three of these operations are not all that different.

Theorem 14.2 The three groups $A + B$, $A \times B$, and B^A are isomorphic.

It is easy to show that $A + B \cong A \times B$. To see that $A + B \cong B^A$, we define the map $f: B^A \to A + B$ by $f(\alpha; \beta) = \alpha^{-1}\beta$, and verify that f is an isomorphism. Note that these three operations are commutative; in fact, $A + B \equiv B + A$, $A \times B \equiv B \times A$, and $B^A \cong A^B$.

Table 14.2 introduces notation for five well-known permutation groups of degree p. In these terms, we can describe the groups of two familiar graphs with p points.

Theorem 14.3 a) The group $\Gamma(G)$ is S_p if and only if $G = K_p$ or $G = \bar{K}_p$.

b) If G is a cycle of length p, then $\Gamma(G) \equiv D_p$.

Thus two particular permutation groups of degree p, namely S_p and D_p, belong to graphs with p points. For all $p \geq 6$, there exists an identity graph with p points and in fact whenever $p \geq 7$, there is an identity tree.

THE GROUP OF A COMPOSITE GRAPH

Now we are ready to study the group associated with a graph formed from other graphs by various operations. Since every automorphism of a graph preserves both adjacency and nonadjacency, an obvious but important result immediately follows.

Theorem 14.4 A graph and its complement have the same group,

$$\Gamma(\bar{G}) = \Gamma(G). \tag{14.5}$$

A "composite graph" is the result of one or more operations on disjoint graphs. The group of a composite graph may often be expressed in terms of the groups of the constituent graphs. Frucht [F10] described the group of a graph nG which consists of n disjoint copies of a connected graph G.

Theorem 14.5 If G is a connected graph, then

$$\Gamma(nG) = S_n[\Gamma(G)]. \tag{14.6}$$

To illustrate the theorem, consider the graph $G = 5K_3$, whose group is $S_5[S_3]$. An automorphism of G can always be obtained by performing an arbitrary automorphism on each of the five triangles, and then following this by any permutation of the triangles among themselves.

Theorem 14.6 If G_1 and G_2 are disjoint, connected, nonisomorphic graphs, then

$$\Gamma(G_1 \cup G_2) \equiv \Gamma(G_1) + \Gamma(G_2). \tag{14.7}$$

Any graph G can be written as $G = n_1G_1 \cup n_2G_2 \cup \cdots \cup n_rG_r$, where n_i is the number of components of G isomorphic to G_i. Applying the last two theorems, we have the result,

$$\Gamma(G) \equiv S_{n_1}[\Gamma(G_1)] + S_{n_2}[\Gamma(G_2)] + \cdots + S_{n_r}[\Gamma(G_r)]. \tag{14.8}$$

Corollary 14.6(a) The group of the union of two graphs is the sum of their groups,

$$\Gamma(G_1 \cup G_2) \equiv \Gamma(G_1) + \Gamma(G_2), \tag{14.9}$$

if and only if no component of G_1 is isomorphic with a component of G_2.

The next corollary follows from Theorem 14.4, the preceding corollary, and the fact that the complement of the join of two graphs is the union of their complements, that is,

$$\overline{G_1 + G_2} = \bar{G}_1 \cup \bar{G}_2. \tag{14.10}$$

Corollary 14.6(b) The group of the join of two graphs is the sum of their groups,

$$\Gamma(G_1 + G_2) \equiv \Gamma(G_1) + \Gamma(G_2), \tag{14.11}$$

if and only if no component of $\bar{G}_1$ is isomorphic with a component of $\bar{G}_2$.

A nontrivial graph G is *prime* if $G = G_1 \times G_2$ implies that G_1 or G_2 is trivial; G is *composite* if it is not prime. Sabidussi [S5] observed that the cartesian product of graphs is commutative and associative. He also developed a criterion for the group of the product of two graphs to be the product of their groups. Since he proved that every nontrivial graph is the unique product of prime graphs, the meaning of relatively prime graphs is clear.

Theorem 14.7 The group of the product of two graphs is the product of their groups,

$$\Gamma(G_1 \times G_2) \equiv \Gamma(G_1) \times \Gamma(G_2), \tag{14.12}$$

if and only if G_1 and G_2 are relatively prime.

Sabidussi [S4] settled the question raised in [H21] by providing a criterion for the group of the lexicographic product (composition) of two

Table 14.3

THE GROUPS OF THE LITTLE CONNECTED GRAPHS

Graph	Group	Graph	Group
•	S_1		$S_2 + S_2$
•——•	S_2		$S_2 + E_2$
	S_3		D_4
	$E_1 + S_2$		$S_2[E_2]$
	S_4		$E_1 + S_3$

graphs to be the composition of their groups. The *neighborhood* of a point u is the set $N(u)$ consisting of all points v which are adjacent with u. The *closed neighborhood* is $N[u] = N(u) \cup \{u\}$.

Theorem 14.8 If G_1 is not totally disconnected, then the group of the composition of two graphs G_1 and G_2 is the composition of their groups,

$$\Gamma(G_1[G_2]) \equiv \Gamma(G_1)[\Gamma(G_2)], \tag{14.13}$$

if and only if the following two conditions hold:

1. If there are two points in G_1 with the same neighborhood, then G_2 is connected.

2. If there are two points in G_1 with the same closed neighborhood, then $\bar{G}_2$ is connected.

With these results, the groups of all graphs with $p \leq 4$ points can be symbolized. The group of one of these graphs, namely $K_4 - x$, has already been illustrated. The groups of the disconnected graphs are not given in Table 14.3 but can be obtained by using Theorem 14.4.

The conditions for the group of the lexicographic products of two graphs to be identical to the composition of their groups are rather complex. This suggests that another operation on graphs be constructed for the purpose of realizing the composition of their groups only up to group isomorphism.

The *corona* $G_1 \circ G_2$ of two graphs G_1 and G_2 was defined by Frucht and Harary [FH1] as the graph G obtained by taking one copy of G_1 (which has

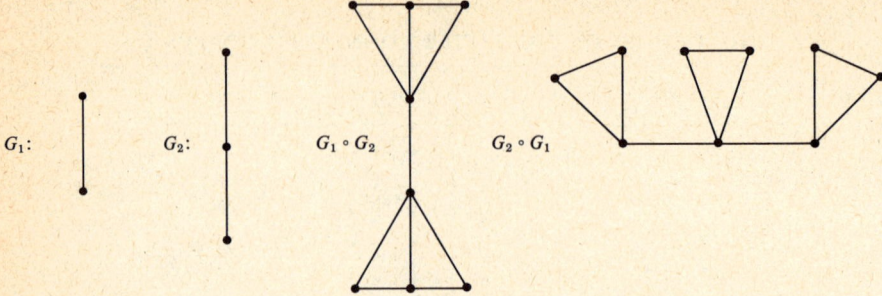

Fig. 14.3. Two graphs and their two coronas.

p_1 points) and p_1 copies of G_2, and then joining the ith point of G_1 to every point in the ith copy of G_2. For the graphs $G_1 = K_2$ and $G_2 = K_{1,2}$, the two different coronas $G_1 \circ G_2$ and $G_2 \circ G_1$ are shown in Fig. 14.3. It follows from the definition of the corona that $G_1 \circ G_2$ has $p_1(1 + p_2)$ points and $q_1 + p_1 q_2 + p_1 p_2$ lines.

Theorem 14.9 The group of the corona of two graphs G_1 and G_2 can be written explicitly in terms of the composition of their groups,

$$\Gamma(G_1 \circ G_2) \equiv \Gamma(G_1)[E_1 + \Gamma(G_2)], \tag{14.14}$$

if and only if G_1 or $\bar{G}_2$ has no isolated points.

The term E_1 in (14.14) when applied to Corollary 14.6(a) gives the next result.

Corollary 14.9(a) The group of the corona $G_1 \circ G_2$ of two graphs is isomorphic to the composition $\Gamma(G_1)[\Gamma(G_2)]$ of their groups if and only if G_1 or $\bar{G}_2$ has no isolated points.

GRAPHS WITH A GIVEN GROUP

König [K10, p. 5] asked: When is a given abstract group isomorphic with the group of some graph? An affirmative answer to this question was given constructively by Frucht [F8]. His proof that every group is the group of some graph makes use of the Cayley "color-graph of a group" [C4] which we now define. Let $F \doteq \{f_0, f_1, \cdots, f_{n-1}\}$ be a finite group of order n whose identity element is f_0. Let each nonidentity element f_i in F have associated with it a different color. The *color-graph* of F, denoted $D(F)$, is a complete symmetric digraph whose points are the n elements of F. In addition, each arc of $D(F)$, say from f_i to f_j, is labeled with the color associated with the element $f_i^{-1} f_j$ of F. Of course, in practice we simply label both points and arcs of $D(F)$ with the elements of F.

For example, consider the cyclic group of order 3, $C_3 = \{0, 1, 2\}$. The color-graph $D(C_3)$ is shown in Fig. 14.4.

Frucht observed the next result, which is simple but very useful.

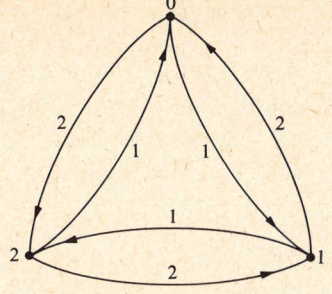

Fig. 14.4. The color graph of the cyclic group C_3.

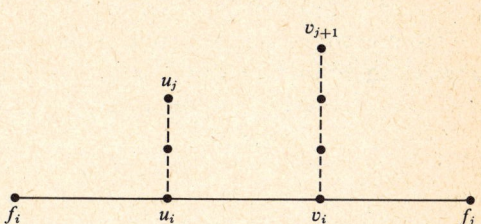

Fig. 14.5. Doubly-rooted graph to replace arc $f_i f_j$.

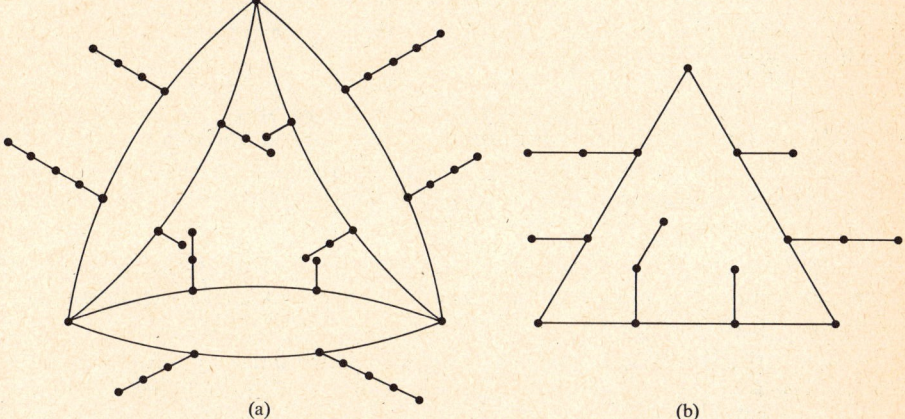

(a)   (b)

Fig. 14.6. Frucht's graph whose group is C_3 and a smaller such graph.

Lemma 14.10(a) Every finite group F is isomorphic with the group of those automorphisms of $D(F)$ which preserve arc colors.

To construct a graph G whose group $\Gamma(G)$ is isomorphic with F, Frucht replaced each arc $f_i f_j$ in $D(F)$ by a doubly rooted graph. This is done in such a way that every arc of the same color is replaced by the same graph. We show in Fig. 14.5 the graph which replaces the arc $f_i f_j$. Let $f_i^{-1} f_j = f_k$ and introduce new points $\{u_m\}$ and $\{v_m\}$ so that in Fig. 14.5 the paths joining u_i with u_j and v_i with v_{j+1} contain $2k - 2$ and $2k - 1$ points respectively. In effect, Frucht's construction assigns a colorful undirected arrow to each arc $f_i f_j$. Thus the resulting graph G has $n^2(2n - 1)$ points and $\Gamma(G) \cong F$.

Theorem 14.10 For every finite abstract group F, there exists a graph G such that $\Gamma(G)$ and F are isomorphic.

The graph obtained by this method from the cyclic group C_3 is shown in Fig. 14.6(a). It should be clear from this example that the number of points in any graph so constructed is excessive. Graphs with a given group and fewer points can be obtained when the group is known to have $m < n$ generators. In that case the color-graph is modified to include only directed lines which correspond to the m generators. Thus a graph containing $n(m + 1)(2m + 1)$ points can be obtained for the given group. Since C_3 can be generated by one element, there is a graph with 18 points for C_3. It is shown in Fig. 14.6(b).

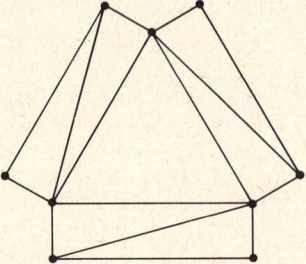

Fig. 14.7. The smallest graph whose group is C_3.

The inefficiency of even this improvement of the method of construction is shown by the graph of Fig. 14.7. This is the unique smallest graph whose automorphism group is cyclic of order three [HP3] and it has only 9 points and 15 lines.

Later Frucht [F9] showed that one could also specify that G be cubic. It was becoming apparent that requiring G to have a given abstract group of automorphisms was not a severe restriction. In fact Sabidussi [S2] showed that there are many graphs with a given abstract group having one of several other specified properties such as connectivity, chromatic number, and degree of regularity.

Theorem 14.11 Given any finite, abstract, nontrivial group F and an integer j $(1 \leq j \leq 4)$, there are infinitely many nonhomeomorphic graphs G such that G is connected, has no point fixed by every automorphism, $\Gamma(G) \cong F$, and G also has the property P_j, defined by

P_1: $\kappa(G) = n,$ $\quad n \geq 1$

P_2: $\chi(G) = n,$ $\quad n \geq 2$

P_3: G is regular of degree $n,$ $\quad n \geq 3$

P_4: G is spanned by a subgraph homeomorphic to a given graph.

When Theorem 14.11 was published, Izbicki [I1] looked into the problem of constructing a graph with a given group which satisfies several of these conditions simultaneously. By exploiting the results of Sabidussi [S2] on the product of two graphs and making some constructions, he was able to obtain a corresponding result involving regular graphs of arbitrary degree and chromatic number.

Corollary 14.11(a) Given any finite group F and integers n and m where $n \geq 3$ and $2 \leq m \leq n$, there are an infinite number of graphs G such that $\Gamma(G) \cong F$, $\chi(G) = m$, and G is regular of degree n.

SYMMETRIC GRAPHS

The study of symmetry in graphs was initiated by Foster [F6], who made a tabulation of symmetric cubic graphs. Two points u and v of the graph G are *similar* if for some automorphism α of G, $\alpha(u) = v$. A *fixed point* is not similar to any other point. Two lines $x_1 = u_1v_1$ and $x_2 = u_2v_2$ are called *similar* if there is an automorphism α of G such that $\alpha(\{u_1, v_1\}) = \{u_2, v_2\}$. We consider only graphs with no isolated points. A graph is *point-symmetric* if every pair of points are similar; it is *line-symmetric* if every pair of lines are similar; and it is *symmetric* if it is both point-symmetric and line-symmetric. The smallest graphs that are point-symmetric but not line-symmetric (the triangular prism $K_3 \times K_2$) and vice versa (the star $K_{1,2}$) are shown in Fig. 14.8.

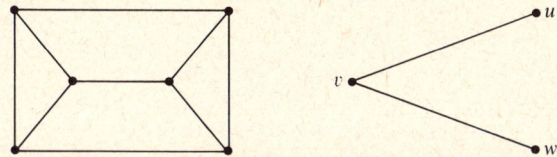

Fig. 14.8. A point-symmetric and a line-symmetric graph.

Note that if α is an automorphism of G, then it is clear that $G - u$ and $G - \alpha(u)$ are isomorphic. Therefore, if u and v are similar, then $G - u \cong G - v$. Surprisingly, the converse of this statement is not true.* The graph in Fig. 14.9 provides a counterexample. It is the smallest graph which has dissimilar points u and v such that $G - u \cong G - v$, see [HP5].

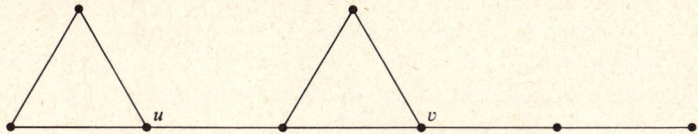

Fig. 14.9. A counterexample to a conjecture.

The *degree of a line* $x = uv$ is the unordered pair (d_1, d_2) with $d_1 = \deg u$, and $d_2 = \deg v$. A graph is *line-regular* if all lines have the same degree. In Fig. 14.10, the complete bipartite graph $K_{2,3}$ is shown; it is line-symmetric but not point-symmetric and is line-regular of degree $(2, 3)$.

---

* A purported proof of Ulam's conjecture depended heavily on this converse.

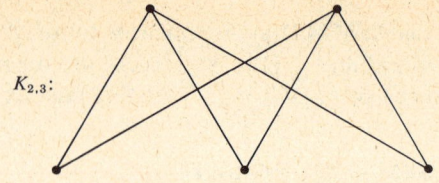

$K_{2,3}$:

Fig. 14.10. A line-regular line-symmetric
graph.

We next state a theorem due to Elayne Dauber whose corollaries describe properties of line-symmetric graphs. Note the obvious but important observation that every line-symmetric graph is line-regular.

Theorem 14.12 Every line-symmetric graph with no isolated points is point-symmetric or bipartite.

Proof. Consider a line-symmetric graph G with no isolated points, having q lines. Then for any line x, there are at least q automorphisms $\alpha_1, \alpha_2, \cdots, \alpha_q$ of G which map x onto the lines of G. Let $x = v_1 v_2$, $V_1 = \{\alpha_1(v_1), \cdots, \alpha_q(v_1)\}$, and $V_2 = \{\alpha_1(v_2), \cdots, \alpha_q(v_2)\}$. Since G has no isolated points, the union of V_1 and V_2 is V. There are two possibilities: V_1 and V_2 are disjoint or they are not.

CASE 1. If V_1 and V_2 are disjoint, then G is bipartite.

Consider any two points u_1 and w_1 in V_1. If they are adjacent, then there is a line y joining them. Hence for some automorphism α_i, we have $\alpha_i(x) = y$. This implies that one of these two points is in V_1 and the other is in V_2, a contradiction. Hence V_1 and V_2 constitute a partition of V such that no line joins two points in the same subset. By definition, G is bipartite.

CASE 2. If V_1 and V_2 are not disjoint, then G is point-symmetric.

Let u and w be any two points of G. We wish to show that u and w are similar. If u and w are both in the same set, say V_1, then there exists automorphism α with $\alpha(v_1) = u$ and β with $\beta(v_1) = w$. Thus $\beta\alpha^{-1}(u) = w$ so that any two points u and w in the same subset are similar. If u is in V_1 and w is in V_2, let v be a point in both V_1 and V_2. Since v is similar with u and with w, u and w are similar to each other.

Corollary 14.12(a) If G is line-symmetric and the degree of every line is (d_1, d_2) with $d_1 \neq d_2$, then G is bipartite.

Corollary 14.12(b) If a graph G with no isolated points is line-symmetric, has an odd number of points, and the degree of every line is (d_1, d_2) with $d_1 = d_2$, then G is point-symmetric.

Corollary 14.12(c) If G is line-symmetric, has an even number of points, and is regular of degree $d \geq p/2$, then G is point-symmetric.

With these three corollaries, the only line-symmetric graphs not yet characterized are those having an even number of points which are regular of degree $d < p/2$. The polygon with six points is an example of such a line-symmetric graph which is both point-symmetric and bipartite. The icosahedron, the dodecahedron, and the Petersen graph are examples of such line-symmetric graphs which are point-symmetric but not bipartite. But not all regular line-symmetric graphs are point-symmetric, as Folkman [F5] discovered.

Theorem 14.13 Whenever $p \geq 20$ is divisible by 4, there exists a regular graph G with p points which is line-symmetric but not point-symmetric.

HIGHLY SYMMETRIC GRAPHS

Following Tutte [T20], an *n-route* is a walk of length n with specified initial point in which no line succeeds itself. A graph G is *n-transitive*, $n \geq 1$, if it has an *n*-route and if there is always an automorphism of G sending each *n*-route onto any other *n*-route. Obviously a cycle of any length is *n*-transitive for all n, and a path of length n is *n*-transitive. Note that not every line-symmetric graph is 1-transitive. For example, in the line-symmetric graph $K_{1,2}$ of Fig. 14.8, there is no automorphism sending the 1-route uv onto the 1-route vw.

If W is an *n*-route $v_0 v_1 \cdots v_n$ and u is any point other than v_{n-1} adjacent with v_n, then the *n*-route $v_1 \cdots v_n u$ is called a *successor* of W. If W terminates in an endpoint of G, then obviously W has no successor. For this reason, it is specified in the next two theorems that G is a graph with no endpoints. We now have a sufficient condition [T20, p. 60] for *n*-transitivity.

Theorem 14.14 Let G be a connected graph with no endpoints. If W is an *n*-route such that there is an automorphism of G from W onto each of its successors, then G is *n*-transitive.

There is a straightforward relationship [T20, p. 61] between *n*-transitivity and the girth of a graph.

Theorem 14.15 If G is connected, *n*-transitive, is not a cycle, has no endpoints and has girth g, then $n \leq 1 + g/2$.

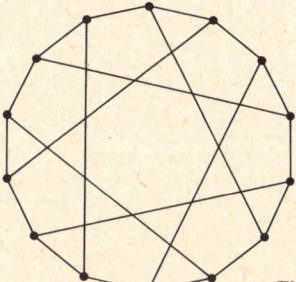

Fig. 14.11. The Heawood graph.

Using Theorem 14.14, it can be shown that the Heawood graph in Fig. 14.11 is 4-transitive. Furthermore, it is easily seen from Theorem 14.15 that this graph is not 5-transitive.

There are regular graphs called "cages" which are, in a sense, even more highly symmetric than n-transitive graphs. A graph G is *n-unitransitive** if it is connected, cubic, and n-transitive, and if for any two n-routes W_1 and W_2, there is exactly one automorphism α of G such that $\alpha W_1 = W_2$. An *n-cage*, $n \geq 3$, is a cubic graph of girth n with the minimum possible number of points. Information about cages is presented in the next statement [T20, pp. 71–83].

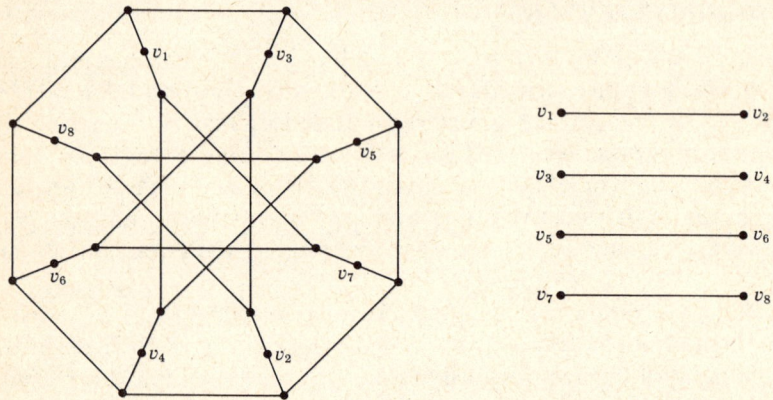

Fig. 14.12. The 7-cage is the union of the above subgraphs as labeled.

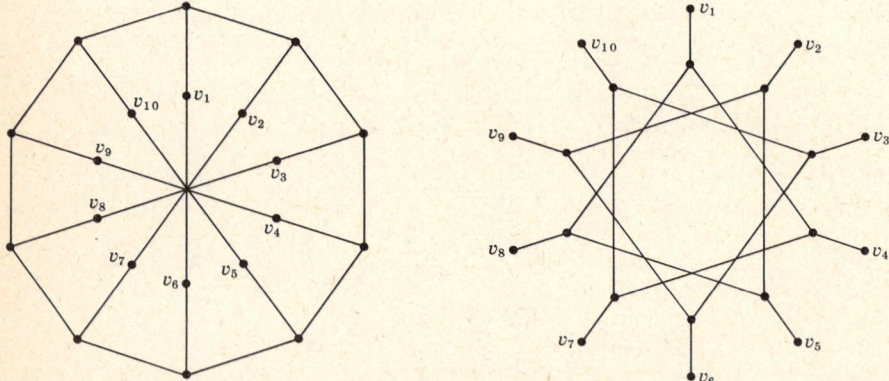

Fig. 14.13. The 8-cage is the union of the above subgraphs as labeled.

---

* Called *n-regular* in [T20, p. 62].

Table 14.4

THE KNOWN CAGES

n	The n-cage	n	The n-cage
3	K_4 (shown in Fig. 2.1)	6	Heawood graph (Fig. 14.11)
4	$K_{3,3}$ (Fig. 2.5)	7	McGee graph (Fig. 14.12)
5	Petersen graph (Fig. 9.6)	8	Levi graph (Fig. 14.13)

Theorem 14.16 There exists an n-cage for all $n \geq 3$. For $n = 3$ to 8 there is a unique n-cage. Each of these n-cages is t-unitransitive for some $t = t(n)$, namely, $t(3) = 2$, $t(4) = t(5) = 3$, $t(6) = t(7) = 4$, and $t(8) = 5$.

All the known cages are now specified.

There are no n-transitive cubic graphs for $n > 5$, hence no n-unitransitive ones; see Tutte [T8]. However, there are other n-unitransitive graphs, $n \leq 5$, in addition to the cages. In particular, Frucht [F11] constructed a 1-unitransitive graph of girth 12 with 432 points, the cube Q_3 and the dodecahedron (Fig. 1.5) are 2-unitransitive, and Coxeter [C10] found 3-unitransitive graphs other than the 4-cage and 5-cage. One of these is shown in Fig. 14.14.

This graph is a member of a class of graphs defined in [CH3]. For any permutation α in S_p, the α-*permutation graph* of a labeled graph G is the union of two disjoint copies G_1 and G_2 of G together with the lines joining point v_i of G_1 with $v_{\alpha(i)}$ of G_2. Thus Fig. 14.14 shows a permutation graph of the cycle C_{10}. The front cover of this book shows all four permutation graphs of C_5.

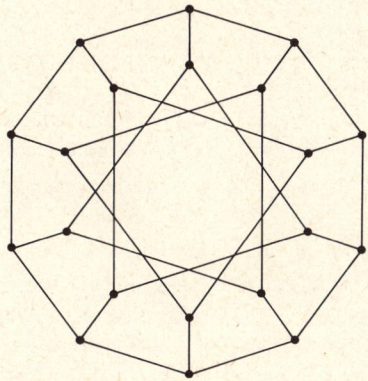

Fig. 14.14. Another 3-unitransitive graph.

EXERCISES

14.1 Find the groups of the following graphs: (a) $\overline{3K_2}$, (b) $\overline{K}_2 + C_4$, (c) $K_{m,n}$, (d) $K_{1,2}[K_2]$, (e) $K_4 \cup C_4$.

14.2 If a connected graph G has a point which is not in a cycle of length four, then G is prime.   (Sabidussi [S2])

14.3 Let G be connected with $p > 3$. Then $L(G)$ is prime if and only if G is not $K_{m,n}$ for $m, n \geq 2$.   (Palmer [P1])

14.5 Construct a connected graph with 11 points whose group is cyclic of order 6.

14.6 Construct a graph with 14 points whose group is cyclic of order 7.

(Sabidussi [S3])

*14.7 Let $c(m)$ be the smallest number of points in a graph whose group is isomorphic to C_m. Then the values of $c(m)$ for $m = n^r$ and n prime are

   a) $c(2) = 2$, and $c(2^r) = 2^r + 6$ when $r > 1$.
   b) $c(n^r) = n^r + 2n$ for $n = 3, 5$.
   c) $c(n^r) = n^r + n$ for $n \geq 7$.

[Note: $c(m)$ can be calculated when m is not a prime power, but the expression is complicated.]   (R. L. Meriwether)

14.8 There are no nontrivial identity graphs with less than 6 points.

14.9 There are no cubic identity graphs with less than 12 points.

14.10 Construct a cubic graph whose group is cyclic of order 3.

14.11 The group of the Petersen graph is identical to the line-group of K_5.

14.12 There exists a graph G whose group is the dihedral group D_p such that G is not a cycle or its complement. What is the smallest value of p for which this holds?

14.13 For $p \geq 3$ there are no graphs G such that $\Gamma(G) \equiv A_p$ or C_p. And when $p \geq 4$ there are no digraphs D with $\Gamma(D) \equiv A_p$.   (Kagno [K1], Harary and Palmer [HP10])

14.14 The only connected graph with group isomorphic to S_n, $n \geq 3$,

   a) with n points is K_n,
   b) with $n + 1$ points is $K_{1,n}$,
   c) with $n + 2$ points is $K_1 + \overline{K}_{1,n}$.   (Gewirtz and Quintas [GQ1])

14.15 Given a finite group F, let $G(F)$ be the graph obtained by Frucht's Theorem. Then every nonidentity automorphism of $G(F)$ leaves no point fixed.

14.16 What is the smallest tree T containing dissimilar points u and v such that $T - u \cong T - v$?   (Harary and Palmer [HP2])

14.17 Every connected, point-symmetric graph G is a block.

14.18 A starred polygon is a graph G containing a spanning cycle $v_1 v_2 \cdots v_p v_1$ such that whenever the line $v_m v_{m+n}$ is in G, so are all lines $v_i v_j$ where $j - i \equiv n \pmod{p}$. A connected graph with a prime number p of points is point-symmetric if and only if it is a starred polygon.   (Turner [T4])

14.19 Prove or disprove the following eight statements: If two graphs are point-symmetric (line-symmetric), then so are their join, product, composition, and corona.

14.20 Every symmetric, connected graph of odd degree is 1-transitive.

(Tutte [T20, p. 59])

14.21 Every symmetric, connected, cubic graph is n-transitive for some n.

(Tutte [T20, p. 63])

14.22 Find necessary and sufficient conditions for the point-group and line-group of a graph to be identical. (Harary and Palmer [HP15])

14.23 If G is connected, then $\Gamma(G) \cong \Gamma(L(G))$ if and only if $G \neq K_2, K_{1,3} + x, K_4 - x$, or K_4. (Whitney [W11])

14.24 If G is point-symmetric, then if $\Gamma(G)$ is abelian, it is a group of the form $S_2 + S_2 + \cdots + S_2$. (McAndrew [M8])

14.25 The only doubly transitive graphical permutation group of degree p is S_p.

14.26 Let A and B be two permutation groups acting on the sets $X = \{x_1, x_2, \cdots, x_d\}$ and Y respectively. The *exponentiation group*, denoted $[B]^A$, acts on the functions Y^X. For each permutation α in A and each sequence of permutations $\beta_1, \beta_2, \cdots, \beta_d$ in B there is a unique permutation $[\alpha; \beta_1, \beta_2, \cdots, \beta_d]$ in $[B]^A$ such that for x_i in X and f in Y^X

$$[\alpha; \beta_1, \beta_2, \cdots, \beta_d] f(x_i) = \beta_i f(\alpha x_i).$$

Then the group of the cube Q_n is $[S_2]^{S_n}$ and the line-group of $K_{n,n}$ is $[S_n]^{S_2}$.

(Harary [H18])

*14.27 There exists a unique, smallest graph of girth 5 which is regular of degree 4. It has 19 points and its group is isomorphic to the dihedral group D_{12}.

(Robertson [R18])

14.28 Let G be a triply connected planar (p, q) graph whose group has order s. Then $4q/s$ is an integer and $s = 4q$ if and only if G is one of the five platonic graphs.

(Weinberg [W8], Harary and Tutte [HT4])

14.29 The group of any tree can be obtained from symmetric groups by the operations of sum and composition. (Pólya [P5, p. 209])

14.30 A collection of $p - 1$ transpositions $(u_1 \, v_1), (u_2 \, v_2), \cdots$ on p objects generates the symmetric group S_p if and only if the graph with p points and the $p - 1$ lines $u_i v_i$ is a tree. (Pólya [P5])

14.31 The α-permutation graph of a labeled 2-connected graph G is planar if and only if G is outerplanar and can be drawn in the plane with a cyclic labeling of its points so that α is in the dihedral group D_p. (Chartrand and Harary [CH3])

*14.32 An *endomorphism* of G is a homomorphism from G into itself. The *semigroup of a graph* is the collection of all its endomorphisms. Every finite semigroup with unit is isomorphic with the semigroup of some graph. (Hedrlin and Pultr [HP23])

*14.33 The smallest nontrivial graph having only the identity endomorphism has 8 points. (Hedrlin and Pultr [HP24])

CHAPTER 15

ENUMERATION

How do I love thee? Let me count the ways.

ELIZABETH BARRETT BROWNING

There is something to be said for regarding enumerative methods in combinatorial analysis as more of an art than a science. With the discovery and development of more general and powerful viewpoints and techniques, it is to be hoped that this situation will become reversed. The pioneers in graphical enumeration theory were Cayley, Redfield, and Pólya. In fact, as noted in [HP11], all graphical enumeration methods in current use were anticipated in the unique paper by Redfield [R8] published in 1927 but unfortunately overlooked.

We begin with the easiest enumeration problems, those for labeled graphs. We then present Pólya's classical enumeration theorem and use it to derive counting series for trees and various other kinds of graphs. Pólya's theorem has been generalized to the Power Group Enumeration Theorem which is useful for certain counting problems where the equivalence classes are determined by two permutation groups. For the sake of completeness, we conclude with lists of both solved and unsolved problems in graphical enumeration.

LABELED GRAPHS

All of the labeled graphs with three points are shown in Fig. 15.1. We see that the 4 different graphs with 3 points become 8 different labeled graphs. To obtain the number of labeled graphs with p points, we need only observe that each of the $\binom{p}{2}$ possible lines is either present or absent.

Theorem 15.1 The number of labeled graphs with p points is $2^{\binom{p}{2}}$.

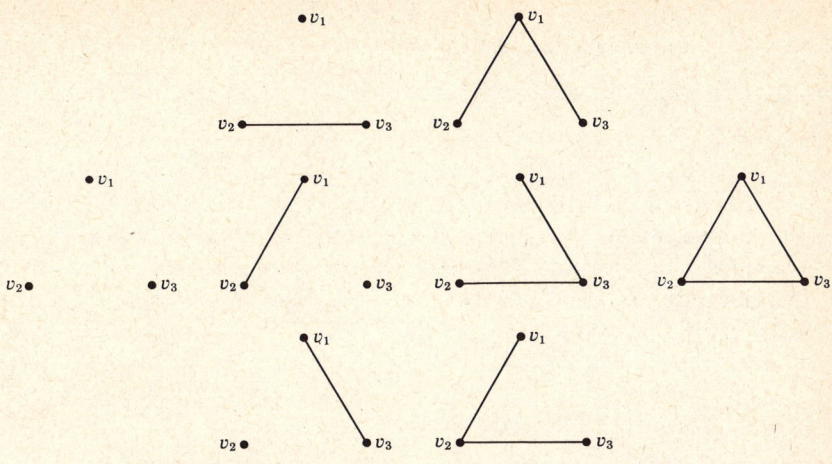

Fig. 15.1. The labeled graphs with three points.

Corollary 15.1(a) The number of labeled (p, q) graphs is

$$\left(\!\!\binom{\binom{p}{2}}{q}\!\!\right).$$

Cayley [C6] was the first to state the corresponding result for trees: The number of labeled trees with p points is p^{p-2}. Since 1889, when Cayley's paper appeared, many different proofs have been found for obtaining his formula. Moon [M15] presents an outline of these various methods of proof, one of which was given in Corollary 13.4(a).

In Fig. 15.2 are all the 16 labeled trees with 4 points. The labels on these trees are understood to be as in the first and last trees shown. We note that among these 16 labeled trees, 12 are isomorphic to the path P_4 and 4 to $K_{1,3}$. The order of $\Gamma(P_4)$ is 2 and that of $\Gamma(K_{1,3})$ is 6. We observe that since $p = 4$ here, we have $12 = 4!/|\Gamma(P_4)|$ and $4 = 4!/|\Gamma(K_{1,3})|$. The

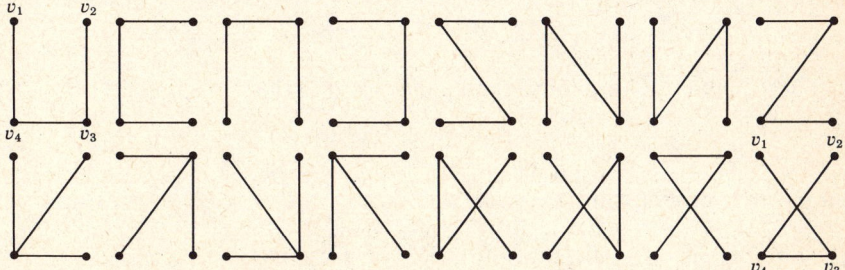

Fig. 15.2. The labeled trees with four points.

expected generalization of these two observations holds not only for trees, but also for graphs, digraphs, relations, and so forth; see [HR1] and [HPR1].

Theorem 15.2 The number of ways in which a given graph G can be labeled is $p!/|\Gamma(G)|$.

Outline of proof. Let A be a permutation group acting on the set X of objects. For any element x in X, the *orbit* of x, denoted $\theta(x)$, is the subset of X which consists of all elements y in X such that for some permutation α in A, $\alpha x = y$. The *stabilizer* of x, denoted $A(x)$, is the subgroup of A which consists of all the permutations in A which leave x fixed. The result follows from an application of the well-known formula $|\theta(x)| \cdot |A(x)| = |A|$ and its interpretation in the present context.

PÓLYA'S ENUMERATION THEOREM

Many enumeration problems are formulated in such a way that the answer can be given by finding a formula for the number of orbits (transitivity systems) determined by a permutation group. Often, weights are assigned to the orbits and Pólya [P5] showed how to obtain a formula which enumerates the orbits according to weight and which depends on the cycle structure of the permutations in the given group. Pólya's theorem in turn depends on a generalization of a well-known counting formula due to Burnside [B20, p. 191].

Theorem 15.3 Let A be a permutation group acting on set X with orbits $\theta_1, \theta_2, \cdots, \theta_n$, and let w be a function which assigns a weight to each orbit. Furthermore, w is defined on X so that $w(x) = w(\theta_i)$ whenever $x \in \theta_i$. Then the sum of the weights of the orbits is given by

$$|A| \sum_{i=1}^{n} w(\theta_i) = \sum_{\alpha \in A} \sum_{x = \alpha x} w(x). \tag{15.1}$$

Proof. We have already seen that the order $|A|$ of the group A is the product $|A(x)| \cdot |\theta(x)|$ for any x in X, where $A(x)$ is the stabilizer of x. Also, since the weight function is constant on the elements in a given orbit, we see that

$$|\theta_i| \, w(\theta_i) = \sum_{x \in \theta_i} w(x),$$

for each orbit θ_i. Combining these facts, we find that

$$|A| \, w(\theta_i) = \sum_{x \in \theta_i} |A(x)| \, w(x).$$

Summing over all orbits, we have

$$|A| \sum_{i=1}^{n} w(\theta_i) = \sum_{i=1}^{n} \sum_{x \in \theta_i} |A(x)| \, w(x),$$

from which (15.1) follows readily.

The conventional form of Burnside's Lemma can now be stated as a corollary to this theorem. For a permutation α, expressed as a product of disjoint cycles, let $j_k(\alpha)$ denote the number of cycles of length k.

Corollary 15.3(a) (Burnside's Lemma) The number $N(A)$ of orbits of the permutation group A is given by

$$N(A) = \frac{1}{|A|} \sum_{\alpha \in A} j_1(\alpha).$$

Let A be a permutation group of order m and degree d. The *cycle index* $Z(A)$ is the polynomial in d variables $a_1, a_2, \cdots, a_d$ given by the formula

$$Z(A) = \frac{1}{|A|} \sum_{\alpha \in A} \prod_{k=1}^{d} a_k^{j_k(\alpha)}. \tag{15.2}$$

Since, for any permutation α, the numbers $j_k = j_k(\alpha)$ satisfy

$$1j_1 + 2j_2 + \cdots + dj_d = d,$$

they constitute a partition of the integer d. It is useful to employ the vector notation $(\mathbf{j}) = (j_1, j_2, \cdots, j_d)$ in describing α. We note that this method of expressing partitions differs from that used in Chapter 6; for example, the partition $5 = 3 + 1 + 1$ corresponds to the vector $(\mathbf{j}) = (2, 0, 1, 0, 0)$.

The classical counting problems to which Pólya's Theorem applies all have the same general form. Let there be given a domain D, a range R, and a weight function w defined on R. To illustrate with a particular weight function, let w assign to each $r \in R$ an ordered pair $w(r) = (w_1 r, w_2 r)$ of nonnegative integers. The objects to be counted will then appear as functions from D to R. To complete the statement of the problem, we need to stipulate when two functions in R^D are considered the same. This is done by specifying a group A which acts on D, so that two functions are equivalent when they are in the same orbit of E^A, where E is the identity group of degree $|R|$.

We digress for a moment to illustrate these ideas with the "necklace problem." Consider necklaces which are to have say 4 beads, some red and some blue. Two such necklaces are regarded as equivalent if they can be made "congruent," with preservation of the colors of their beads. Here the domain D is the set of locations where the beads are to be put, the range R is the set {red bead, blue bead}, and a function $f \in R^D$ is an assignment of one bead to each place, giving a necklace. In this example, A is the dihedral group D_4, and the weight function w can be taken as $w(\text{red bead}) = (1, 0)$ and $w(\text{blue bead} = (0, 1)$.

Following the intuitive terminology of Pólya, domain elements are *places*, range elements are *figures*, functions are *configurations*, and the permutation group A is the *configuration group*. We assign a weight $W(f)$ to each $f \in R^D$ by the equation

$$W(f) = \prod_{d \in D} x^{w_1 f(d)} y^{w_2 f(d)}. \tag{15.3}$$

It is easy to see that each function in a given orbit of R^D under E^A has the same weight, so that the weight of an orbit can be defined as the weight of any function in it.

Suppose there are c_{mn} figures of weight (m, n) in R and C_{mn} orbits (equivalence classes of configurations) of weight $x^m y^n$ in R^D. The *figure counting series*

$$c(x, y) = \sum c_{mn} x^m y^n \qquad (15.4)$$

enumerates the elements of R by weight, and the *configuration counting series*

$$C(x, y) = \sum C_{mn} x^m y^n \qquad (15.5)$$

is the generating function for equivalence classes of functions. Pólya's Theorem [P5] expresses $C(x, y)$ in terms of $c(x, y)$.

If in (15.2) we write $Z(A) = Z(A; a_1, a_2, \cdots, a_d)$, then for any function $h(x, y)$, we define

$$Z(A, h(x, y)) = Z(A; h(x, y), h(x^2, y^2), \cdots, h(x^d, y^d)). \qquad (15.6)$$

Theorem 15.4 (Pólya's Enumeration Theorem) The configuration counting series is obtained by substituting the figure counting series into the cycle index of the configuration group,

$$C(x, y) = Z(A, c(x, y)). \qquad (15.7)$$

Proof. Let α be a permutation in A, and let $\tilde{\alpha}$ be the corresponding permutation in the power group E^A. Assume first that f is a configuration fixed by $\tilde{\alpha}$ and that ζ is a cycle of length k in the disjoint–cycle decomposition of α. Then $f(b) = f(\zeta b)$ for every element b in the representation of ζ, so that all elements permuted by ζ must have the same image under f. Conversely, if the elements of each cycle of the permutation α have the same image under a configuration f, then $\tilde{\alpha}$ fixes f. Therefore, all configurations fixed by $\tilde{\alpha}$ are obtained by independently selecting an element r in R for each cycle ζ of α and setting $f(b) = r$ for all b permuted by ζ. Then if the weight $w(r)$ is (m, n) where $m = w_1 r$ and $n = w_2 r$ and ζ has length k, the cycle ζ contributes a factor of $\sum_{r \in R} (x^m y^n)^k$ to the sum $\sum_{f = \tilde{\alpha}f} W(f)$. Therefore, since

$$\sum_{r \in R} (x^m y^n)^k = c(x^k, y^k),$$

we have, for each α in A,

$$\sum_{f = \tilde{\alpha}f} W(f) = \prod_{k=1}^{s} c(x^k, y^k)^{j_k(\alpha)}.$$

Summing both sides of this equation over all permutations α in A (or equivalently over all $\tilde{\alpha}$ in E^A) and dividing both sides by $|A| = |E^A|$, we obtain

$$\frac{1}{|E^A|} \sum_{\tilde{\alpha} \in E^A} \sum_{f = \tilde{\alpha}f} W(f) = \frac{1}{|A|} \sum_{\alpha \in A} \prod_{k=1}^{s} c(x^k, y^k)^{j_k(\alpha)}. \qquad (15.8)$$

The right hand side of this equation is $Z(A, c(x, y))$. To see that the left hand side is $C(x, y)$, we apply the version of Burnside's lemma given in Theorem 15.3. First note that for the power group E^A, the sum of the weights of the orbits is given by

$$\sum_{i=1}^{n} w(\theta_i) = \sum C_{mn}x^m y^n = C(x, y). \tag{15.9}$$

But it follows at once from (15.1) that the left sides of (15.9) and (15.8) are equal, so that $Z(A, c(x, y)) = C(x, y)$, proving the theorem.

Returning to the necklace problem with four beads mentioned above, we note that the cycle index of the dihedral group D_4 is

$$Z(D_4) = \tfrac{1}{8}(a_1^4 + 2a_1^2 a_2 + 3a_2^2 + 2a_4) \tag{15.10}$$

and the figure counting series is $c(x, y) = x^1 y^0 + x^0 y^1 = x + y$. Substituting $x + y$ into (15.10) in accordance with (15.6), we obtain

$$
\begin{aligned}
Z(D_4, x + y) &= \tfrac{1}{8}\{(x + y)^4 + 2(x + y)^2(x^2 + y^2) \\
&\quad + 3(x^2 + y^2)^2 + 2(x^4 + y^4)\} \\
&= x^4 + x^3 y + 2x^2 y^2 + xy^3 + y^4. \tag{15.11}
\end{aligned}
$$

The coefficient of $x^m y^n$ in (15.11) is the number of different necklaces with four beads, m red and n blue. The 6 different necklaces are shown in Fig. 15.3.

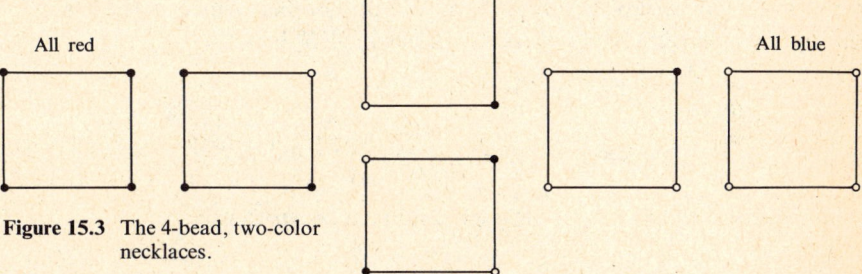

All red   All blue

Figure 15.3 The 4-bead, two-color necklaces.

Incidentally, necklaces can also be counted by using $1 + x$ as the figure counting series instead of $x + y$. In this case a red bead has weight 1 and a blue bead weight 0. Then in $Z(D_4, 1 + x) = x^4 + x^3 + 2x^2 + x + 1$, the coefficient of x^m is the number of necklaces with m red beads and hence $4 - m$ blue ones; compare (15.11). As we shall see in the next section, the figure counting series $1 + x$ plays an important role in enumeration problems, since x^0 indicates absence of a figure and x^1 presence. The reason is indicated in the following consequence [H31] of Pólya's Theorem. An *n-subset* of a set X is a subset with exactly n elements.

Corollary 15.4(a) If A is a permutation group acting on X, then the number of orbits of n-subsets of X induced by A is the coefficient of x^n in $Z(A, 1 + x)$.

In applications of Pólya's Enumeration Theorem, certain permutation groups occur frequently. The formulas for the cycle indexes of the five important permutation groups listed in Table 14.2 are now given. In (15.12) and (15.13), the sum is over all partitions (j) of p. In (15.14), $\phi(k)$ is the "Euler ϕ-function," the number of positive integers less than k and relatively prime to k, with $\phi(1) = 1$.

$$Z(S_p) = \frac{1}{p!} \sum_{(j)} \frac{p!}{\prod_{k=1}^{p} k^{j_k} j_k!} a_1^{j_1} a_2^{j_2} \cdots a_p^{j_p} \tag{15.12}$$

$$Z(A_p) = \frac{1}{p!} \sum_{(j)} \frac{p! \left[1 + (-1)^{j_2 + j_4 + \cdots}\right]}{\prod_{k=1}^{p} k^{j_k} j_k!} a_1^{j_1} a_2^{j_2} \cdots a_p^{j_p} \tag{15.13}$$

$$Z(C_p) = \frac{1}{p} \sum_{k|p} \phi(k) a_k^{p/k} \tag{15.14}$$

$$Z(D_p) = \frac{1}{2} Z(C_p) + \begin{cases} \frac{1}{2} a_1 a_2^{(p-1)/2}, & p \text{ odd} \\ \frac{1}{4}(a_2^{p/2} + a_1^2 a_2^{(p-2)/2}), & p \text{ even} \end{cases} \tag{15.15}$$

$$Z(E_p) = a_1^p \tag{15.16}$$

There are several very useful formulas which give the cycle indexes of the binary operations of the sum, product, composition, and power group of A and B in terms of $Z(A)$ and $Z(B)$. They are given in equations (15.17)–(15.22) and appear in [H31]. By $Z(A)[Z(B)]$ we mean the polynomial obtained by replacing each variable a_k in $Z(A)$ by the polynomial which is the result of multiplying the subscripts of the variables in $Z(B)$ by k.

$$Z(A + B) = Z(A)Z(B). \tag{15.17}$$

$$Z(A \times B) = \frac{1}{|A|} \frac{1}{|B|} \sum_{(\alpha,\beta)} \prod_{r,s=1}^{d,e} a_{m(r,s)}^{d(r,s)j_r(\alpha)j_s(\beta)} \tag{15.18}$$

where $d(r, s)$ and $m(r, s)$ are the g.c.d. and l.c.m. respectively.

$$Z(A[B]) = Z(A)[Z(B)]. \tag{15.19}$$

$$Z(B^A) = \frac{1}{|A| \cdot |B|} \sum_{(\alpha;\beta)} a_k^{j_k(\alpha;\beta)} \tag{15.20}$$

where $(\alpha; \beta) = \beta^\alpha$ and

$$j_1(\alpha; \beta) = \prod_{k=1}^{d} \left(\sum_{s|k} s j_s(\beta)\right)^{j_k(\alpha)} \tag{15.21}$$

and for $k > 1$

$$j_k(\alpha; \beta) = \frac{1}{k} \sum_{s|k} \mu\left(\frac{k}{s}\right) j_1(\alpha^s; \beta^s) \tag{15.22}$$

with μ the familiar number-theoretic möbius function.*

---

* By definition $\mu(n) = 0$ unless n is the product of distinct primes $p_1, \cdots, p_m$ in which case $\mu(n) = (-1)^m$.

ENUMERATION OF GRAPHS

We now describe how to obtain the polynomial $g_p(x)$ which enumerates graphs with a given number p of points. Let g_{pq} be the number of (p, q) graphs and let

$$g_p(x) = \sum_q g_{pq} x^q. \tag{15.23}$$

By inspection of all graphs with 4 points, one easily verifies that

$$g_4(x) = 1 + x + 2x^2 + 3x^3 + 2x^4 + x^5 + x^6. \tag{15.24}$$

Let $V = \{1, 2, \cdots, p\}$ and let $R = \{0, 1\}$. We denote by $D = V^{(2)}$ the collection of subsets $\{i, j\}$ of distinct elements of V, that is, of 2-subsets of V. Then each function f from D into R represents a graph whose p points are the elements of V, in which i is adjacent with j whenever $f\{i, j\} = 1$. Thus the image of $\{i, j\}$ under f is 1 or 0 in accordance with the presence or absence of a line joining i and j. The weight function w on R is defined by $w(0) = 0$ and $w(1) = 1$, so that it is the identity function. Hence the figure counting series is $c(x) = 1 + x$. Specializing (15.3) to one variable, the weight of a function f is given by

$$W(f) = x \sum w(f\{i, j\}) \tag{15.25}$$

where the sum is taken over all pairs $\{i, j\}$ in $V^{(2)}$. Thus the weight of function f is the number of lines in the graph corresponding to f.

Now let E_2 be the identity group acting on R and let S_p act on V. We denote by $S_p^{(2)}$ the *pair group* which acts on $V^{(2)}$ whose permutations are induced by S_p. That is, for each permutation α in S_p, there is a permutation α' in $S_p^{(2)}$ such that $\alpha'\{i, j\} = \{\alpha i, \alpha j\}$. Applying Pólya's theorem to the configuration group $S_p^{(2)}$, we have the next result, also due to Pólya; see [H11].

Theorem 15.5 The counting polynomial for graphs with p points is

$$g_p(x) = Z(S_p^{(2)}, 1 + x), \tag{15.26}$$

where

$$Z(S_p^{(2)}) = \frac{1}{p!} \sum_{(j)} \frac{p!}{\prod_{k=1}^p j_k! \, k^{j_k}} \prod_{k=1}^{[p/2]} (a_k a_{2k}^{k-1})^{j_{2k}}. \tag{15.27}$$

$$\prod_{k=0}^{[(p-1)/2]} a_{2k+1}^{kj_{2k+1}} \prod_{k=1}^{[p/2]} a_k^{k\binom{j_k}{2}} \prod_{1 \le r < s \le p-1} a_{m(r,s)}^{d(r,s)j_r j_s}.$$

A derivation of (15.27) is also given in [H31, p. 38]. In Appendix I, the number of (p, q)-graphs is tabulated through $p = 9$.

Similar counting formulas have been obtained which enumerate rooted graphs and connected graphs. Various classes of graphs have also been enumerated by modifications of this method. These include directed graphs, pseudographs, and multigraphs. We illustrate some of these enumeration formulas by describing how they follow readily from the preceding theorem. First to enumerate rooted graphs, it is necessary to fix the root point and regard the remaining $p - 1$ points as interchangeable before forming the pair group.

Corollary 15.5(a) The counting polynomial for rooted graphs with p points is

$$r_p(x) = Z((S_1 + S_{p-1})^{(2)}, 1 + x). \tag{15.28}$$

When there are at most two lines joining each pair of points, we need only replace the figure counting series for graphs by $1 + x + x^2$.

Corollary 15.5(b) The counting polynomial for multigraphs with at most two lines joining each pair of points is

$$g_p''(x) = Z(S_p^{(2)}, 1 + x + x^2). \tag{15.29}$$

For arbitrary multigraphs, the figure counting series becomes

$$1 + x + x^2 + x^3 + \cdots = \frac{1}{1 - x}.$$

Corollary 15.5(c) The counting polynomial for multigraphs with p points is

$$m_p(x) = Z\left(S_p^{(2)}, \frac{1}{1 - x}\right). \tag{15.30}$$

The enumeration of digraphs [H11] is also accomplished, as for graphs, by finding a formula for the cycle index of the appropriate configuration group and applying Pólya's theorem. For digraphs, we need to use the *reduced ordered pair group*, denoted $S_p^{[2]}$. As before S_p acts on $V = \{1, 2, \cdots, p\}$. By definition, $S_p^{[2]}$ acts on $V^{[2]}$, the ordered pairs of distinct elements of V, as induced by S_p. Thus every permutation α in S_p induces a permutation α' in $S_p^{[2]}$ such that $\alpha'(i, j) = (\alpha i, \alpha j)$ for (i, j) in $V^{[2]}$. Applying Pólya's theorem to the cycle index of $S_p^{[2]}$, obtain $d_p(x)$, the polynomial in which the coefficient of x^q is the number of digraphs with q directed lines.

Theorem 15.6 The counting polynomial for digraphs with p points is

$$d_p(x) = Z(S_p^{[2]}, 1 + x), \tag{15.31}$$

where

$$Z(S_p^{[2]}) = \frac{1}{p!} \sum_{(j)} \frac{p!}{\prod_{k=1}^p j_k! \, k^{j_k}} \prod_{k=1}^p a_k^{(k-1)j_k + 2k\binom{j_k}{2}} \prod_{1 \le r < s \le p-1} a_{m(r,s)}^{2 j_r j_s \, d(rs)}. \tag{15.32}$$

Of course this theorem has corollaries analogous to those of Theorem 15.5.

Appendix II includes a table for the number of digraphs with $p \leq 8$ points.

Although rooted trees and trees were counted much earlier than graphs, the enumeration of graphs was presented above because of the simplicity of the figure counting series, viz. $1 + x$. We will see that for tree counting purposes, the most useful figure counting series is the generating function for rooted trees themselves.

ENUMERATION OF TREES

In order to find the number of trees it is necessary to start by counting rooted trees. A *rooted tree* has one point, its root, distinguished from the others. Let T_p be the number of rooted trees with p points. From Fig. 15.4 in which the root of each tree is visibly distinguished from the other points, we see that $T_4 = 4$. The counting series for rooted trees is denoted by

$$T(x) = \sum_{p=1}^{\infty} T_p x^p. \tag{15.33}$$

We define t_p and $t(x)$ similarly for unrooted trees.

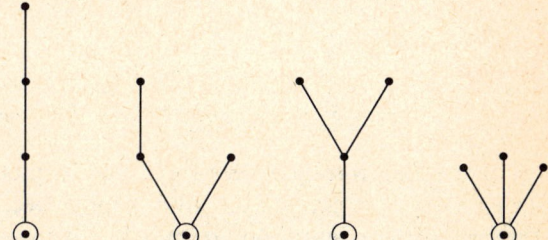

Fig. 15.4. The rooted trees with four points.

A recursive type of expression for counting rooted trees was found by Cayley [C2].

Theorem 15.7 The counting series for rooted trees is given by

$$T(x) = x \prod_{r=1}^{\infty} (1 - x^r)^{-T_r}. \tag{15.34}$$

It is possible to convert (15.34) into a form expressing $T(x)$ in terms of itself by taking the logarithm of both sides and then manipulating power series appropriately. This leads to (15.35), a result first obtained by Pólya [P5] by exploiting his enumeration theorem.

Theorem 15.8 The counting series for rooted trees satisfies the functional equation

$$T(x) = x \exp \sum_{r=1}^{\infty} \frac{1}{r} T(x^r). \tag{15.35}$$

Proof. Let $T^{(n)}(x)$ be the generating function for those rooted trees in which the root has degree n, so that

$$T(x) = \sum_{n=0}^{\infty} T^{(n)}(x). \tag{15.36}$$

Thus for example, $T^{(0)}(x) = x$ counts the rooted trivial graph, while the *planted trees* (rooted at an endpoint) are counted by $T^{(1)}(x) = xT(x)$. In general a rooted tree with root degree n can be regarded as a configuration whose figures are the n rooted trees obtained on removing the root. Figure 15.5 illustrates this for $n = 3$.

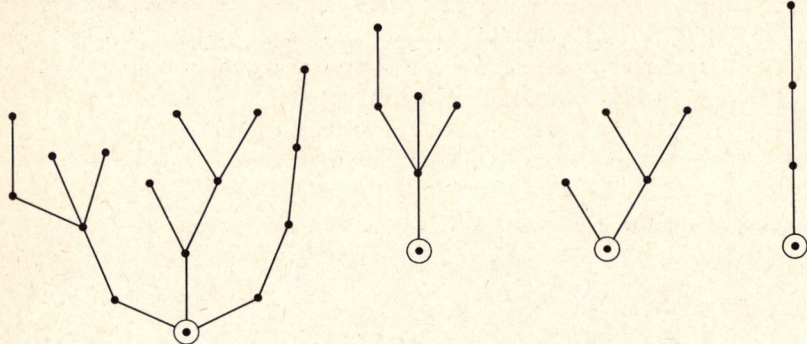

Fig. 15.5. A given rooted tree T and its constituent rooted trees.

Since these n rooted trees are mutually interchangeable without altering the isomorphism class of the given rooted tree, the figure counting series is $T(x)$ and the configuration group is S_n, giving

$$T^{(n)}(x) = xZ(S_n, T(x)). \tag{15.37}$$

The factor x accounts for the removal of the root of the given tree since the weight of a tree is the number of points.

Fortunately, there is a well-known and easily derived identity which may now be invoked (where $Z(S_0)$ is defined as 1):

$$\sum_{n=0}^{\infty} Z(S_n, h(x)) = \exp \sum_{r=1}^{\infty} \frac{1}{r} h(x^r). \tag{15.38}$$

On combining the last three equations, we obtain (15.35).

Cayley [C5] was the first to derive an expression for t_p in terms of the numbers T_n with $n < p$. He did this by counting separately the number of centered and bicentered trees. Pólya [P5] obtained an alternate expression for t_p by considering separately trees with 1 and 2 centroid points. Otter [O8] discovered the neatest possible formula for the number of trees in terms of the number of rooted trees, entirely by means of generating functions.

Actually, Otter's equation (15.41) can be derived directly from the Cayley or Pólya expressions for t_p, as shown in [H12], by repeated application of the adage, "Whenever you see two consecutive summation signs, interchange the order of summation." Otter derived (15.41) from the next observation, which is of independent interest; it is sometimes called "the dissimilarity characteristic equation for trees." A *symmetry line* joins two similar points.

Theorem 15.9 For any tree T, let p^* and q^* be the number of similarity classes of points and lines, respectively, and let s be the number of symmetry lines. Then $s = 0$ or 1 and

$$p^* - (q^* - s) = 1. \qquad (15.39)$$

Outline of proof. Whenever T has one central point or two dissimilar central points, there is no symmetry line, so $s = 0$. In this case there is a subtree of T which contains exactly one point from each similarity class of points in T and exactly one line from each class of lines. Since this subtree has p^* points and q^* lines, we have $p^* - q^* = 1$.

The other possibility is that T has two similar central points and hence $s = 1$. In this case there is a subtree which contains exactly one point from each similarity class of points in T and, except for the symmetry line, one line from each class of lines. Therefore this subtree has p^* points and $q^* - 1$ lines and so $p^* - (q^* - 1) = 1$. Thus in both cases (15.39) holds.

We also require a special theorem of Pólya [P5] which was designed for counting 1–1 functions. For convenience we use $Z(A_n - S_n)$ as an abbreviation for $Z(A_n) - Z(S_n)$.

Theorem 15.10 The configuration counting series $C(x)$ for 1–1 functions from a set of n interchangeable elements into a set with figure counting series $c(x)$ is obtained by substituting $c(x)$ into $Z(A_n - S_n)$:

$$C(x) = Z(A_n - S_n, c(x)). \qquad (15.40)$$

Although we will only use (15.40) in the case $n = 2$, it provides a useful enumeration device in other contexts [HP20], and it enables us to present a very concise proof of Otter's formula for counting trees.

Theorem 15.11 The counting series for trees in terms of rooted trees is given by the equation

$$t(x) = T(x) - \tfrac{1}{2}[T^2(x) - T(x^2)]. \qquad (15.41)$$

Proof. For $i = 1$ to t_n, let p_i^*, q_i^*, and s_i be the numbers of similarity classes of points, lines, and symmetry lines for the ith tree with n points. Since $1 = p_i^* - (q_i^* - s_i)$ for each i, by (15.39), we sum over i to obtain

$$t_n = T_n - \sum_i (q_i^* - s_i). \qquad (15.42)$$

Furthermore $\Sigma (q_i^* - s_i)$ is the number of trees having n points which are rooted at a line, not a symmetry line. Consider a tree T and take any line y of T which is not a symmetry line. Then $T - y$ may be regarded as two rooted trees which must be nonisomorphic. Thus each nonsymmetry line of a tree corresponds to an unordered pair of different rooted trees. Counting these pairs of trees is equivalent to counting 1–1 functions from a set of two interchangeable elements into the collection of rooted trees. Therefore we apply Theorem 15.10 with $T(x)$ as the figure counting series to obtain

$$\sum_{n=1}^{\infty} \left[x^n \sum_{i=1}^{t_n} (q_i^* - s_i) \right] = Z(A_2 - S_2, T(x)). \qquad (15.43)$$

Since $Z(A_2) = a_1^2$ and $Z(S_2) = \frac{1}{2}(a_1^2 + a_2)$, we have

$$Z(A_2 - S_2, T(x)) = \frac{1}{2}[T^2(x) - T(x^2)]. \qquad (15.44)$$

Now the formula in the theorem follows from (15.42)–(15.44).

Using (15.35) and (15.41) we obtain the explicit numbers of rooted and unrooted trees through $p = 12$,

$$T(x) = x + x^2 + 2x^3 + 4x^4 + 9x^5 + 20x^6 + 48x^7$$
$$+ 115x^8 + 286x^9 + 719x^{10} + 1842x^{11} + 4766x^{12} + \cdots$$
$$(15.45)$$

$$t(x) = x + x^2 + x^3 + 2x^4 + 3x^5 + 6x^6 + 11x^7 + 23x^8$$
$$+ 47x^9 + 106x^{10} + 235x^{11} + 551x^{12} + \cdots \qquad (15.46)$$

The diagrams for the trees counted in the first 10 terms of (15.46) may be found in Appendix III, along with a table displaying t_p and T_p for $p \le 26$.

The methods used to derive Theorem 15.11 can be extended to count various species of trees. We illustrate with two species, homeomorphically irreducible trees and identity trees [HP20]; others can be handled similarly, for example colored trees [R14], trees with a given partition [HP20], and so on. Let $h(x)$, $H(x)$, and $\bar{H}(x)$ be the counting series for homeomorphically irreducible trees, rooted trees, and planted trees respectively.

Theorem 15.12 Homeomorphically irreducible trees are counted by the three equations,

$$\bar{H}(x) = \frac{x^2}{1 + x} \exp \sum_{r=1}^{\infty} \frac{\bar{H}(x^r)}{rx^r} \qquad (15.47)$$

$$H(x) = \frac{1 + x}{x} \bar{H}(x) - \frac{1}{2x} [\bar{H}^2(x) - \bar{H}(x^2)]. \qquad (15.48)$$

$$h(x) = H(x) - \frac{1}{x^2} [\bar{H}^2(x) - \bar{H}(x^2)]. \qquad (15.49)$$

The number of homeomorphically irreducible trees through 12 points is found to be:

$$h(x) = x + x^2 + x^4 + x^5 + 2x^6 + 2x^7 + 4x^8 + 5x^9$$
$$+ 10x^{10} + 14x^{11} + 26x^{12} + \cdots \quad (15.50)$$

Let $u(x)$ and $U(x)$ be the counting series for identity trees and rooted trees for which the automorphism group is the identity group.

Theorem 15.13 Identity trees are counted by the equations

$$U(x) = x \exp \sum_{n=1}^{\infty} (-1)^{n+1} \frac{U(x^n)}{n}. \quad (15.51)$$

$$u(x) = U(x) - \tfrac{1}{2}[U^2(x) + U(x^2)]. \quad (15.52)$$

The number of identity trees through 12 points is given by

$$u(x) = x + x^7 + x^8 + 3x^9 + 6x^{10} + 15x^{11} + 29x^{12} + \cdots \quad (15.53)$$

POWER GROUP ENUMERATION THEOREM

There is a class of enumeration problems which can be solved using a power group as the configuration group. Consider the power group B^A acting on R^D. The number of configurations (equivalence classes of functions determined by B^A) can be derived from Pólya's Theorem as shown in [HP8], and was discovered by deBruijn [B18] and [B19] in another formulation. The equation (15.54) given by the next theorem can be readily modified to count functions with respect to their weights.

Theorem 15.14 (Power Group Enumeration Theorem) The number of equivalence classes of functions in R^D determined by the power group B^A is

$$N(B^A) = \frac{1}{|B|} \sum_{\beta \in B} Z(A; m_1(\beta), m_2(\beta), \cdots, m_d(\beta)) \quad (15.54)$$

where

$$m_k(\beta) = \sum_{s|k} s j_s(\beta). \quad (15.55)$$

To illustrate, we consider once again the necklace problem illustrated in Fig. 15.3, but here we allow the two colors a, b of beads (say red and blue) to be interchangeable. Clearly the number of necklaces with 4 beads of two interchangeable colors is $N(S_2^{D_4})$, the number of orbits of the power group $S_2^{D_4}$. For the identity permutation $(a)(b)$ of S_2 we have from (15.55)

$$m_k((a)(b)) = 2$$

for all k. For the transposition (ab) in S_2, $m_k((ab))$ is 0 or 2 according as k is odd or even. Applying (15.54) we see that the number of necklaces with interchangeable colors is

$$\tfrac{1}{2}[Z(D_4; 2, 2, 2, 2) + Z(D_4; 0, 2, 0, 2)].$$

By substitution in formula (15.10) for $Z(D_4)$ we find that the number of such necklaces is 4. This calculation is easily verified by observing that the last two necklaces of Fig. 15.3 are equivalent to the first two, when red and blue are interchangeable.

The self-complementary graphs with 4 and 5 points are shown in Fig. 2.13. The result of Read [R5] for the number s_p of self-complementary graphs with p points is easily obtained from the Power Group Enumeration Theorem. For this purpose we define a new equivalence relation $\sim$ for graphs with p points, namely $G_1 \sim G_2$ if $G_1 \cong G_2$ or $G_1 \cong \bar{G}_2$. Let c_p be the number of such equivalence classes of graphs with p points. Since we are dealing with graphs on p points, we take $A = S_p^{(2)}$ acting on $D^{(2)}$. Because a graph and its complement are equivalent we let $B = S_2$ act on $R = \{0, 1\}$. Then under the power group B^A, two functions f_1 and f_2 from $D^{(2)}$ into R are equivalent whenever they represent the same graph or one represents the complement of the other. We have already seen the result of applying (15.55) to the permutations of S_2. Hence we have

$$c_p = \tfrac{1}{2}[Z(S_p^{(2)}; 2, 2, 2, 2, \cdots) + Z(S_p^{(2)}; 0, 2, 0, 2, \cdots)]. \qquad (15.56)$$

But since $s_p = 2c_p - g_p$, we have the following formula obtained by Read.

Theorem 15.15 The number s_p of self-complementary graphs on p points is

$$s_p = Z(S_p^{(2)}; 0, 2, 0, 2, \cdots). \qquad (15.57)$$

Finite automata have also been counted using the Power Group Enumeration Theorem by Harrison [H34] and Harary and Palmer [HP12]. The groups for this problem are subgroups of the product of two power groups.

SOLVED AND UNSOLVED GRAPHICAL ENUMERATION PROBLEMS

There have now been three lists of unsolved graphical enumeration problems in the literature, [H24], [H30], and most recently [H32, p. 30]. It is frequently necessary to bring these lists up to date. Because of the fact that new problems arise as old ones are solved, the number of unsolved problems remains relatively constant. We might note that it is extremely unlikely that all of these enumeration problems will soon be settled. For included among such solutions there would be enough information to decide the validity of the Four Color Conjecture by comparing the number of planar graphs with the number of 4-colorable planar graphs.

Table 15.1 presents the fourth list of unsolved graphical enumeration problems and is so titled. All of these problems can, of course, be proposed

Table 15.1

UNSOLVED GRAPHICAL ENUMERATION PROBLEMS IV

Category	Enumerate
Digraph	Strong digraphs Unilateral digraphs Digraphs with a source Transitive digraphs Digraphs which are both self-complementary and self-converse
Traversability	Hamiltonian graphs Hamiltonian cycles in a given graph Eulerian trails in a given graph
Topological	Simplicial complexes k-colorable graphs Planar k-colorable graphs Rooted planar graphs Edge-rooted plane maps Plane and planar graphs
Symmetry	Symmetric graphs Identity graphs Graphs with given automorphism group
Applications	Even subgraphs of a labeled 3-lattice Even subgraphs of a labeled 2-lattice with given area Even subgraphs of a given labeled graph Pavings of a 2-lattice Animals
Miscellaneous	Line graphs Latin squares Graphs with given radius or diameter Graphs with given girth or circumference Graphs with given connectivity Graphs with given genus, thickness, chromatic number, etc.

for labeled graphs as well, and several of them have been solved in the labeled case. A few additional definitions are needed for understanding these problems, each of which challenges the mathematician to determine the number of configurations named in terms of suitable parameters. Definitions needed for the digraph category may be found in the next chapter.

Tutte [T15] studied the enumeration of plane maps rooted in the following way to destroy any symmetry that might be present. An *edge-rooted plane map* is obtained from a plane map by orienting an arbitrary edge and by then designating one of the two faces incident with this edge as the exterior face of the map.

A *2-lattice* is a graph whose points are ordered pairs of integers (i, j) where $i = 0, 1, \cdots, m$ and $j = 0, 1, \cdots, n$; two of these points are adjacent whenever their distance in the cartesian plane is 1. A *3-lattice* is defined similarly. An *even subgraph H* of a graph G is one in which every point has even degree. Thus every even subgraph of a 2-lattice has a certain area, the number of squares contained in its cycles.

By a *paving* of a 2-lattice is meant a covering of the squares of the lattice by a given number of single unit squares and double squares like dominoes. Of course larger and more complicated paving problems can be proposed.

There are three kinds of cell growth problems, one each for the triangle, the square, and the hexagon, the only three regular polygons which can cover the plane. Then an *animal* is a simply connected configuration containing a given number of triangles, squares, or hexagons; see [H32, pp. 33–38].

We include here a comprehensive list of solved problems (which will inevitably be incomplete) in the hope that unnecessary duplication of combinatorial effort will be minimized. References are given to papers where solutions are reported; unpublished solutions are credited only by the name of the (eventual) author. These solved problems (Table 15.2) are divided into four categories: trees, graphs, digraphs, and miscellaneous.

Table 15.2

SOLVED GRAPHICAL ENUMERATION PROBLEMS

Trees	
Trees	Pólya [P5], Otter [O8]
Labeled trees	Cayley [C6], Moon [M15]
Rooted trees	Pólya [P5]
Rooted trees with given height	Riordan [R16]
Endlessly labeled trees	Harary, Mowshowitz, Riordan [HMR1]
Plane trees	Harary, Prins, Tutte [HPT1]
Plane trees with given partition	Tutte [T18], Harary, Tutte [HT2]
Homeomorphically irreducible trees	Harary, Prins [HP20]
Identity trees	Harary, Prins [HP20]
Trees with given partition	Harary, Prins [HP20]
Trees with given group	Prins [P8]
Trees with given diameter	Harary, Prins [HP20]
Directed trees	Harary, Prins [HP20]
Oriented trees	Harary, Prins [HP20]
Signed trees	Harary, Prins [HP20]
Trees of given strength	Harary, Prins [HP20]
Trees of given type	Harary, Prins [HP20]
Block-cutpoint trees	Harary, Prins [HP20]
Colored trees	Riordan [R14]
Forests	Harary, Palmer [HP16]

Graphs

Graphs	Pólya [H11], Davis [D1]
Rooted graphs	Harary [H11]
Line rooted graphs	Harary [H31]
Graphs rooted at an oriented line	Harary, Palmer [HP1]
Connected graphs	Riddell, Uhlenbeck [RU1], Harary [H11]
Multigraphs	Harary [H11]
Graphs of given strength	Harary [H11]
Graphs of given type	Harary [H11]
Spanning subgraphs and supergraphs of G	Harary [H13], [H14], [H19]
Self-complementary graphs	Read [R5]
Signed graphs	Harary [H10], Harary, Palmer [HP13]
Unicyclic graphs	Riordan [R15, p. 147]
Eulerian graphs	(R. W. Robinson)
Graphs with given partition	Parthasarathy [P2]
Pseudographs with given partition	Read [R3]
Superposed graphs	Read [R3]
Superposed graphs with interchangeable colors	Palmer, Robinson [PR1]
Cubic graphs	(R. W. Robinson)
Nonseparable graphs	(R. W. Robinson)
k-colored graphs	Robinson [R19]
Bicolorable graphs	Harary, Prins [HP21]
Edge-rooted triangulated maps	Tutte [T14]
Cacti	Harary, Norman [HN2], Harary, Uhlenbeck [HU1]
Graphs with given blocks	Ford, Norman, Uhlenbeck [FNU1]
Block graphs	Harary, Prins [HP22]

Digraphs

Digraphs	Harary [H11], Davis [D1]
Weakly connected digraphs	Harary [H11]
Self-complementary digraphs	Read [R5]
Self-converse digraphs	Harary, Palmer [HP9]
Oriented graphs	Harary [H16]
Orientations of a given graph	Harary, Palmer [HP4]
Tournaments	Davis [D2]
Strong tournaments	Moon [M16]
Labeled transitive digraphs	Evans, Harary, Lynn [EHL1]
Digraphs with given partition	Harary, Palmer [HP7]
Digraphs with all points of outdegree 2	(C. P. Lawes)
Acyclic digraphs	(R. W. Robinson)
Functional digraphs	Harary [H23], Read [R4]
Eulerian trails in a given digraph	de Bruijn, Ehrenfest [BE1], Smith, Tutte [ST1]

Table 15.2 (*continued*)

Miscellaneous

Automata	Harrison [H34], Harary, Palmer [HP12]
Necklace problems	Harary [H31]
Algebras of various kinds	Harrison [H35]
Boolean functions	Pólya [P6], Slepian [S14]
Labeled series-parallel networks	Carlitz, Riordan [CR1]
Periodic sequences	Gilbert, Riordan [GR1]
Acyclic simplicial complexes	Harary, Palmer [HP17], Beineke, Moon [BM1], Beineke, Pippert [BP1]

EXERCISES

15.1 In how many ways can the graphs (a) $\bar{K}_3 + K_2$, (b) $K_3 \times K_2$, (c) $K_{1,2}[K_2]$ be labeled?

15.2 Write expressions for the cycle indexes of $S_3 + S_2$, $S_3 \times S_2$, $S_3[S_2]$, $S_3^{S_2}$, and $S_2^{S_3}$.

15.3 There is an integer k such that $Z(C_n, 2) = Z(D_n, 2)$ holds for all $n \leq k$ and fails whenever $n > k$. Find k.

15.4 The number of partitions of n into at most m parts is the coefficient of x^n in

$$Z\left(S_m, \frac{1}{1 - x}\right).$$

15.5 Calculate $Z(S_5^{(2)})$ and $g_5(x)$. Verify this result using Appendix I.

15.6 Find a counting series for unicyclic graphs.   (Riordan [R15, p. 147])

15.7 Let $g(x, y) = \Sigma_{p=1}^{\infty} g_p(x)y^p$ be the generating function for graphs and let $c(x, y)$ be that for connected graphs. Then

$$g(x, y) = \exp \sum_{r=1}^{\infty} \frac{1}{r} c(x^r, y^r).$$

[Note the similarity to equation (15.38).]

15.8 Find the number of trees with p points which are (a) planted and labeled, (b) rooted and labeled.

15.9 Let G be a labeled graph obtained from K_p by deleting r independent lines. The number of spanning trees of G is $(p - 2)^r p^{p-2-r}$.   (Weinberg [W7])

15.10 The number of rooted trees satisfies the inequality $T_{n+1} \leq \Sigma_{i=1}^n T_i T_{n-i+1}$. It follows that

$$T_n \leq \frac{1}{n}\binom{2n - 2}{n - 1}.$$ (Otter [O8])

15.11 Define the numbers $R_n^{(i)}$ by the equation $R_n^{(i)} = R_{n-1}^{(i)} + T_{n+1-i}$. Then the number of rooted trees can be found using

$$nT_{n+1} = \sum_{i=1}^{n} iT_i R_n^{(i)}. \qquad \text{(Otter [O8])}$$

15.12 Determine the number s_p of self-complementary graphs for $p = 8$ and 9, both by formula (15.57) and by constructing them.

15.13 Derive a counting formula for self-complementary digraphs.   (Read [R5])

15.14 Let s_p and $\vec{s}_p$ be the numbers of self-complementary graphs and digraphs, respectively. Then $s_{4n} = \vec{s}_{2n}$.   (Read [R5])

15.15 For any permutation group A with cycle index $Z(A)$ as given in (15.2), the number of orbits of A is

$$N(A) = \left. \frac{\partial}{\partial a_1} Z(A) \right|_{\text{all } a_i = 1.}$$

Therefore the number of similarity classes of points in a given graph G (whose permutation group $\Gamma(G)$ has the variables y_i in its cycle index) is

$$p^* = \left. \frac{\partial}{\partial y_1} Z(\Gamma(G)) \right|_{\text{all } y_i = 1.}$$

15.16 Let G be a connected graph with n similarity classes of blocks. If p^* is the number of dissimilar points of G and p_k^* is the number of dissimilar points in blocks of the kth similarity class, then

$$p^* - 1 = \sum_{k=1}^{n} (p_k^* - 1).$$

Prove Theorem 15.9 as a corollary.   (Harary and Norman [HN3])

CHAPTER 16

DIGRAPHS

I shot an arrow in the air,
It fell to earth I know not where.
HENRY WADSWORTH LONGFELLOW

There is so much to digraph theory that it is possible to write an entire book on the subject.* For the most part we shall emphasize in this chapter those properties of digraphs which set them apart from graphs. Thus we begin by developing three different kinds of connectedness: strong, unilateral, and weak. After presenting the Directional Duality Principle, we study matrices related to digraphs and the analogue of the Matrix Tree Theorem for graphs. We close with a brief description of tournaments.

DIGRAPHS AND CONNECTEDNESS

We have already seen all the digraphs with 3 points and 3 arcs in Fig. 2.4. For completeness, we begin with definitions, including a few from Chapter 2. A *digraph D* consists of a finite set V of *points* and a collection of ordered pairs of distinct points. Any such pair (u, v) is called an *arc* or *directed line* and will usually be denoted uv. The arc uv goes from u to v and is *incident* with u and v. We also say that u is adjacent *to* v and v is adjacent *from u.* The *outdegree* od(v) of a point v is the number of points adjacent from it, and the *indegree* id(v) is the number adjacent to it.

A (*directed*) *walk* in a digraph is an alternating sequence of points and arcs, $v_0, x_1, v_1, \cdots, x_n, v_n$ in which each arc x_i is $v_{i-1}v_i$. The *length* of such a walk is n, the number of occurrences of arcs in it. A *closed walk* has the same first and last points, and a *spanning walk* contains all the points. A *path* is a walk in which all points are distinct; a *cycle* is a nontrivial closed walk with all points distinct (except the first and last). If there is a path from

---

* In fact this has been done, [HNC1]. Most of the theorems in this chapter are proved in that book. Also Moon [M16] has written a monograph on tournaments.

u to v, then v is said to be *reachable from u*, and the *distance, $d(u, v)$*, from u to v is the length of any shortest such path.

Each walk is directed from the first point v_0 to the last v_n. We also need a concept which does not have this property of direction and is analogous to a walk in a graph. A *semiwalk* is again an alternating sequence $v_0, x_1, v_1, \cdots, x_n, v_n$ of points and arcs, but each arc x_i may be either $v_{i-1}v_i$ or v_iv_{i-1}. A *semipath, semicycle*, and so forth, are defined as expected.

Whereas a graph is either connected or it is not, there are three different ways in which a digraph may be connected, and each has its own idiosyncrasies. A digraph is *strongly connected*, or *strong*, if every two points are mutually reachable; it is *unilaterally connected*, or *unilateral*, if for any two points at least one is reachable from the other; and it is *weakly connected*, or *weak*, if every two points are joined by a semipath. Clearly, every strong digraph is unilateral and every unilateral digraph is weak, but the converse statements are not true. A digraph is *disconnected* if it is not even weak. We note that the *trivial digraph*, consisting of exactly one point, is (vacuously) strong since it does not contain two distinct points.

We may now state necessary and sufficient conditions for a digraph to satisfy each of the three kinds of connectedness.

Theorem 16.1 A digraph is strong if and only if it has a spanning closed walk, it is unilateral if and only if it has a spanning walk, and it is weak if and only if it has a spanning semiwalk.

Corresponding to connected components of a graph, there are three different kinds of components of a digraph. A *strong component* of a digraph is a maximal strong subgraph; a *unilateral component* is a maximal unilateral subgraph; and a *weak component* is a maximal weak subgraph. It is very easy to verify that every point and every arc of a digraph D is in just one weak component and in at least one unilateral component. Furthermore each point is in exactly one strong component, and an arc lies in one strong component or none, depending on whether or not it is in some cycle.

The strong components of a digraph are the most important among these. One reason is the way in which they yield a new digraph which,

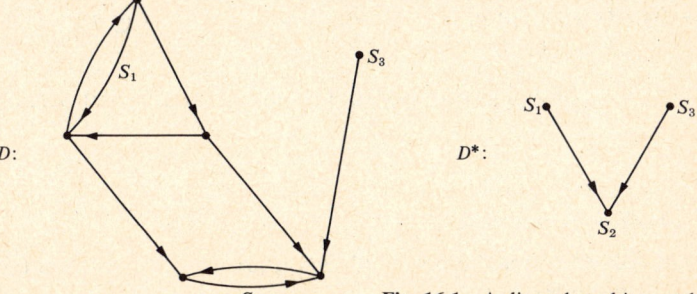

Fig. 16.1. A digraph and its condensation.

although simpler, retains some structural properties of the original. Let $S_1, S_2, \cdots, S_n$ be the strong components of D. The *condensation D^** of D has the strong components of D as its points, with an arc from S_i to S_j whenever there is at least one arc in D from a point of S_i to a point in S_j. (See Fig. 16.1.)

It follows from the maximality of strong components that the condensation D^* of any digraph D has no cycles. Obviously the condensation of any strong digraph is the trivial digraph. It can be shown that a digraph is unilateral if and only if its condensation has a unique spanning path.

DIRECTIONAL DUALITY AND ACYCLIC DIGRAPHS

The *converse* digraph D' of D has the same points as D and the arc uv is in D' if and only if the arc vu is in D. Thus the converse of D is obtained by reversing the direction of every arc of D. We have already encountered other converse concepts, such as indegree and outdegree, and these concepts concerned with direction are related by a rather powerful principle. This is a classical result in the theory of binary relations.

Principle of Directional Duality For each theorem about digraphs, there is a corresponding theorem obtained by replacing every concept by its converse.

We now illustrate how this principle generates new results. An *acyclic* digraph contains no directed cycles.

Theorem 16.2 An acyclic digraph has at least one point of outdegree zero.

Proof. Consider the last point of any maximal path in the digraph. This point can have no points adjacent from it since otherwise there would be a cycle or the path would not be maximal.

The dual theorem follows immediately by applying the Principle of Directional Duality. In keeping with the use of D' to denote the converse of digraph D, we shall use primes to denote dual results.

Theorem 16.2′ An acyclic digraph D has at least one point of indegree zero.

It was noted that the condensation of any digraph is acyclic, and the preceding results give some information about acyclic digraphs. We now provide several characterizations.

Theorem 16.3 The following properties of a digraph D are equivalent.

1. D is acyclic.
2. D^* is isomorphic to D.
3. Every walk of D is a path.
4. It is possible to order the points of D so that the adjacency matrix $A(D)$ is upper triangular.

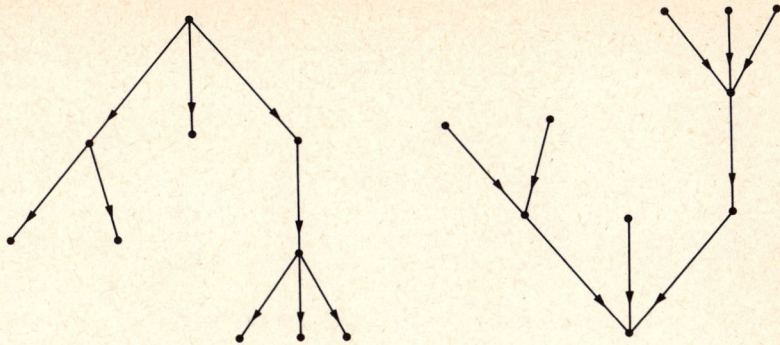

Fig. 16.2. An out-tree and the converse in-tree.

Two dual types of acyclic digraphs are of particular interest. A *source* in D is a point which can reach all others; a *sink* is the dual concept. An *out-tree** is a digraph with a source having no semicycles; an *in-tree* is its dual, see Fig. 16.2.

Theorem 16.4 A weak digraph is an out-tree if and only if exactly one point has indegree 0 and all others have indegree 1.

Theorem 16.4′ A weak digraph is an in-tree if and only if exactly one point has outdegree 0 and all others have outdegree 1.

We next consider some digraphs which are closely related to the above. A *functional* digraph is one in which every point has outdegree 1; a *contra-functional* digraph is dual, see Fig. 16.3. The next theorem and its dual provide structural characterizations.

Theorem 16.5 The following are equivalent for a weak digraph D.

1. D is functional.
2. D has exactly one cycle, the removal of whose arcs results in a digraph in which each weak component is an in-tree with its sink in the cycle.
3. D has exactly one cycle Z, and the removal of any arc of Z results in an in-tree.

A *point basis* of D is a minimal collection of points from which all points are reachable. Thus, a set S of points of a digraph D is a point basis if and only if every point of D is reachable from a point of S and no point of S is reachable from any other.

Theorem 16.6 Every acyclic digraph has a unique point basis consisting of all points of indegree 0.

* This is called an "arborescence" by Berge[B12, p. 13].

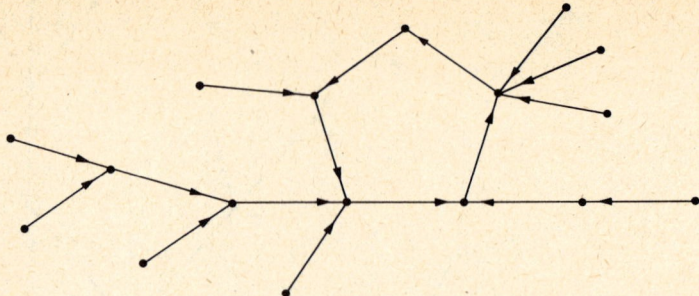

Fig. 16.3. A weak functional digraph.

Corollary 16.6(a) Every point basis of a digraph D consists of exactly one point from each of those strong components in D which form the point basis of D^*.

A *1-basis* is a minimal collection S of mutually nonadjacent points such that every point of D is either in S or adjacent from a point of S. Every digraph has a point basis, but not every digraph has a 1-basis. For example, no odd cycle has one. A criterion for an arbitrary digraph to have a 1-basis has not yet been found. The theorem by Richardson [R9] generalizes its corollary, due to von Neumann and Morgenstern [NM1], and discovered in their study of game theory.

Theorem 16.7 Every digraph with no odd cycles has a 1-basis.

Corollary 16.7(a) Every acyclic digraph has a 1-basis.

DIGRAPHS AND MATRICES

The *adjacency matrix $A(D)$ of a digraph D* is the $p \times p$ matrix $[a_{ij}]$ with $a_{ij} = 1$ if $v_i v_j$ is an arc of D, and 0 otherwise. As the example in Fig. 16.4 shows, the row sums of $A(D)$ give the outdegrees of the points of D and the column sums are the indegrees.

As in the case of graphs, the powers of the adjacency matrix A of a digraph give information about the number of walks from one point to another.

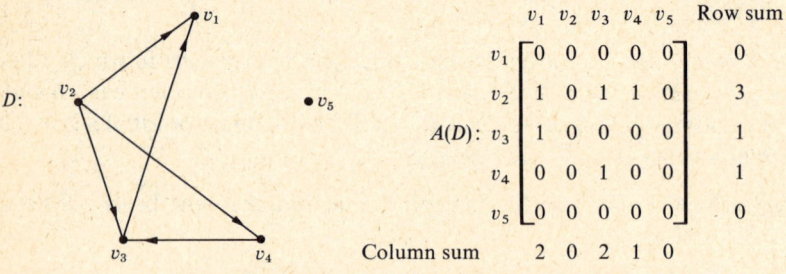

Fig. 16.4. A digraph and its adjacency matrix.

Theorem 16.8 The i, j entry $a_{ij}^{(n)}$ of A^n is the number of walks of length n from v_i to v_j.

We mention briefly three other matrices associated with D, namely the reachability matrix, the distance matrix, and the detour matrix. In R, the *reachability matrix*, r_{ij} is 1 if v_j is reachable from v_i, and 0 otherwise. The i, j entry in the *distance matrix* gives the distance from the point v_i to the point v_j, and is infinity if there is no path from v_i to v_j. In the *detour matrix*, the i, j entry is the length of any longest path from v_i to v_j, and again is infinity if there is no such path. These three matrices for the digraph D of Fig. 16.4 are:

Reachability Matrix	Distance Matrix	Detour Matrix
$\begin{bmatrix} 1 & 0 & 0 & 0 & 0 \\ 1 & 1 & 1 & 1 & 0 \\ 1 & 0 & 1 & 0 & 0 \\ 1 & 0 & 1 & 1 & 0 \\ 0 & 0 & 0 & 0 & 1 \end{bmatrix}$	$\begin{bmatrix} 0 & \infty & \infty & \infty & \infty \\ 1 & 0 & 1 & 1 & \infty \\ 1 & \infty & 0 & \infty & \infty \\ 2 & \infty & 1 & 0 & \infty \\ \infty & \infty & \infty & \infty & 0 \end{bmatrix}$	$\begin{bmatrix} 0 & \infty & \infty & \infty & \infty \\ 3 & 0 & 2 & 1 & \infty \\ 1 & \infty & 0 & \infty & \infty \\ 2 & \infty & 1 & 0 & \infty \\ \infty & \infty & \infty & \infty & 0 \end{bmatrix}$

Corollary 16.8(a) The entries of the reachability and distance matrices can be obtained from the powers of A as follows:

(1) for all i, $r_{ii} = 1$ and $d_{ii} = 0$.
(2) $r_{ij} = 1$ if and only if for some n, $a_{ij}^{(n)} > 0$.
(3) $d(v_i, v_j)$ is the least n (if any) such that $a_{ij}^{(n)} > 0$, and is ∞ otherwise.

There is no efficient method for finding the entries of the detour matrix. This problem is closely related to several other long-standing algorithmic questions of graph theory, such as finding spanning cycles and solving the traveling salesman problem.*

The *elementwise product*** $B \times C$ of two matrices $B = [b_{ij}]$ and $C = [c_{ij}]$ has $b_{ij}c_{ij}$ as its i, j entry. The reachability matrix can be useful in finding strong components.

Corollary 16.8(b) Let v_i be a point of a digraph D. The strong component of D containing v_i is determined by the entries of 1 in the ith row (or column) of the matrix $R \times R^T$.

The number of spanning in-trees in a given digraph was found by Bott and Mayberry [BM2] and proved by Tutte [T9]. To give this result, called

---

* Consider a network N obtained from a strong digraph D by assigning a positive integer (cost) to every arc of D. The traveling salesman problem asks for an algorithm for finding a walk in N whereby the salesman can visit each point and return to the starting point while traversing arcs with a minimum total cost.
** Sometimes called the "Hadamard product."

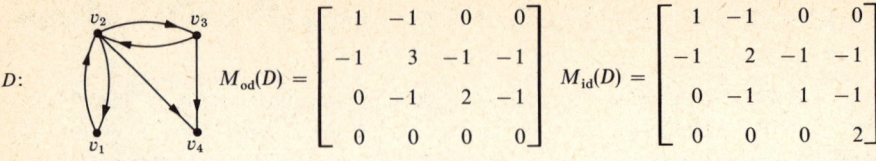

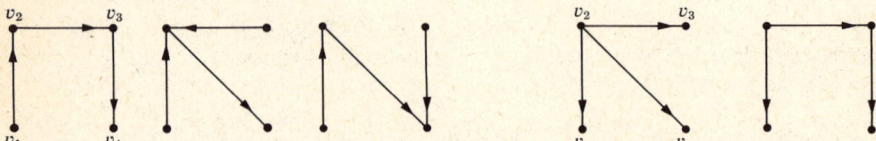

Fig. 16.5. Spanning in-trees and out-trees.

the matrix tree theorem for digraphs, we need some other matrices related to D. Let M_{od} denote the matrix obtained from $-A$ by replacing the ith diagonal entry by od(v_i). The matrix M_{id} is defined dually.

Theorem 16.9 For any labeled digraph D, the value of the cofactor of any entry in the ith row of M_{od} is the number of spanning in-trees with v_i as sink.

Theorem 16.9′ The value of the cofactor of any entry in the jth column of M_{id} is the number of spanning out-trees with v_j as source.

In accordance with Theorem 16.9, the matrix M_{od} of the digraph of Fig. 16.5 has all cofactors of its entries in the fourth row equal to 3 and the three spanning in-trees of D with v_4 as sink are displayed; the directional dual, Theorem 16.9′, is also illustrated by the second column of M_{id} and the two spanning out-trees with v_2 as source.

An *eulerian trail* in a digraph D is a closed spanning walk in which each arc of D occurs exactly once. A digraph is *eulerian* if it has such a trail. Just as in Theorem 7.1 for graphs, one can easily show that a weak digraph D is eulerian if and only if every point of D has equal indegree and outdegree. We will now state a theorem giving the number of eulerian trails in an eulerian digraph. It is sometimes referred to as the BEST theorem after the initials of de Bruijn, van Aardenne-Ehrenfest, Smith, and Tutte; the first two [BE1] and the last two [ST1] discovered the theorem independently. It can be elegantly proved using the matrix tree theorem for digraphs, see Kasteleyn [K4, p. 76].

Corollary 16.9(a) In an eulerian digraph, the number of eulerian trails is

$$c \cdot \prod_{i=1}^{p} (d_i - 1)!$$

where $d_i = \text{id}(v_i)$ and c is the common value of all the cofactors of M_{od}.

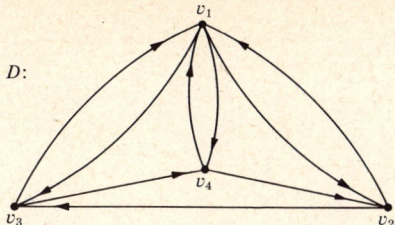

$$M_{od}(D) = \begin{bmatrix} 3 & -1 & -1 & -1 \\ -1 & 2 & -1 & 0 \\ -1 & 0 & 2 & -1 \\ -1 & -1 & 0 & 2 \end{bmatrix}$$

Fig. 16.6. Counting eulerian trails.

Note that for an eulerian digraph D, we have $M_{od} = M_{id}$ and all row sums as well as column sums are zero, so that all cofactors are equal. For the digraph in Fig. 16.6, $c = 7$ and there are 14 eulerian trails. Two of them are $v_1v_2v_3v_4v_2v_1v_3v_1v_4v_1$ and $v_1v_2v_1v_4v_2v_3v_4v_1v_3v_1$.

We have just given some indication of how matrices are used in the study of digraphs. On the other hand digraphs can be used to give information about matrices. Any square matrix $M = [m_{ij}]$ gives rise to a digraph D, and also possibly to loops if arc v_iv_j is in D whenever $m_{ij} \neq 0$. The following algorithm [H25] sometimes simplifies the determination of the eigenvalues and the inverse (if it exists) of a matrix M.

1. Form the digraph D associated with M.
2. Determine the strong components of D.
3. Form the condensation D^*.
4. Order the strong components so that the adjacency matrix of D^* is upper triangular.
5. Reorder the points of D by strong components so that its adjacency matrix A is upper block triangular.
6. Replace each unit entry of A by the entry of M to which it corresponds.

The eigenvalues of M are the eigenvalues of the diagonal blocks of the new matrix, and the inverse of M can be found from the inverses of these diagonal blocks.

When M is a sparse matrix,* (or rather has zero entries strategically located so that there are several strong components), this method can be quite effective. A generalization to a sometimes more powerful but also more involved algorithm using bipartite graphs is given by Dulmage and Mendelsohn [DM2].

TOURNAMENTS

A *tournament* is an oriented complete graph. All tournaments with two, three, and four points are shown in Fig. 16.7. The first with three points is called a *transitive triple*, the second a *cyclic triple*.

---

* In the literature, a sparse matrix has been defined as one with many zeros.

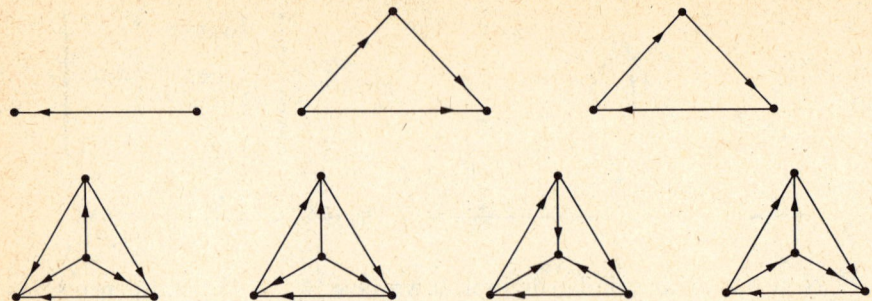

Fig. 16.7. Small tournaments.

In a round-robin tournament, a given collection of players or teams play a game in which the rules of the game do not allow for a draw. Every pair of players encounter each other and exactly one from each pair emerges victorious. The players are represented by points and for each pair of points an arc is drawn from the winner to the loser, resulting in a tournament.

The first theorem on tournaments ever found is due to Rédei [R7]; for small tournaments, it can be verified using Fig. 16.7.

Theorem 16.10 Every tournament has a spanning path.

Proof. The proof is by induction on the number of points. Every tournament with 2, 3, or 4 points has a spanning path, by inspection. Assume the result is true for all tournaments with n points, and consider a tournament T with $n + 1$ points. Let v_0 be any point of T. Then $T - v_0$ is a tournament with n points, so it has a spanning path P, say $v_1 v_2 \cdots v_n$. Either arc $v_0 v_1$ or arc $v_1 v_0$ is in T. If $v_0 v_1$ is in T, then $v_0 v_1 v_2 \cdots v_n$ is a spanning path of T. If $v_1 v_0$ is in T, let v_i be the first point of P for which the arc $v_0 v_i$ is in T, if any. Then $v_{i-1} v_0$ is in T, so that $v_1 v_2 \cdots v_{i-1} v_0 v_i \cdots v_n$ is a spanning path. If no such point v_i exists, then $v_1 v_2 \cdots v_n v_0$ is a spanning path. In any case, we have shown that T has a spanning path, completing the proof.

Szele [S16] extended this result by proving that every tournament has an odd number of spanning paths. Another type of extension of Rédei's theorem was provided by Gallai and Milgram [GM1] who showed that every oriented graph D contains a collection of at most $\beta_0(D)$ point-disjoint paths which cover $V(D)$.

The next theorem is due to Moser [HM2]; its corollary was discovered by Foulkes [F7] and Camion [C1] and is the analogue for strong tournaments of the preceding theorem for arbitrary tournaments.

Theorem 16.11 Every strong tournament with p points has a cycle of length n, for $n = 3, 4, \cdots, p$.

Proof. This proof is also by induction, but on the length of cycles. If a tournament T is strong, then it must have a cyclic triple. Assume that T

has a cycle $Z = v_1 v_2 \cdots v_n v_1$ of length $n < p$. We will show that it has a cycle of length $n + 1$. There are two cases: either there is a point u not in Z both adjacent to and adjacent from points of Z, or there is no such point.

CASE 1.   Assume there is a point u not in Z and points v and w in Z such that arcs uv and wu are in T. Without loss of generality, we assume that arc $v_1 u$ is in T. Let v_i be the first point, going around Z from v_1, for which arc uv_i is in T. Then $v_{i-1}u$ is in T, and $v_1 v_2 \cdots v_{i-1} u v_i \cdots v_n v_1$ is a cycle of length $n + 1$.

CASE 2.   There is no such point u as in Case 1. Hence, all points of T which are not in Z are partitioned into the two subsets U and W, where U is the set of all points adjacent to every point of Z and W is the set adjacent from every point of Z. Clearly these sets are disjoint, and neither set is empty since otherwise T would not be strong. Furthermore, there are points u in U and w in W such that arc wu is in T. Therefore $u v_1 v_2 \cdots v_{n-1} w u$ is a cycle of length $n + 1$ in T.

Hence, there is a cycle of length $n + 1$, completing the proof.

Corollary 16.11(a) A tournament is strong if and only if it has a spanning cycle.

Using terminology from round-robin tournaments, we say that the *score* of a point in a tournament is its outdegree. The next theorem due to Landau [L1] was actually discovered during an empirical study of tournaments (so-called "pecking orders") in which the points were hens and the arcs indicated pecking.

Theorem 16.12 The distance from a point with maximum score to any other point is 1 or 2.

The number of transitive triples can be given in terms of the scores of the points; see Harary and Moser [HM2]. As a corollary, one can readily obtain the well-known formula of Kendall and Smith [KS1], which has proved useful in statistical analysis. It was generalized from cyclic triples to larger strong subtournaments by Beineke and Harary [BH4].

Theorem 16.13 The number of transitive triples in a tournament with score sequence $(s_1, s_2, \cdots, s_p)$ is $\Sigma\, s_i(s_i - 1)/2$.

Corollary 16.13(a) The maximum number of cyclic triples among all tournaments with p points is

$$
t(p, 3) = \begin{cases} \dfrac{p^3 - p}{24} & \text{if } p \text{ is odd,} \\[2ex] \dfrac{p^3 - 4p}{24} & \text{if } p \text{ is even.} \end{cases}
$$

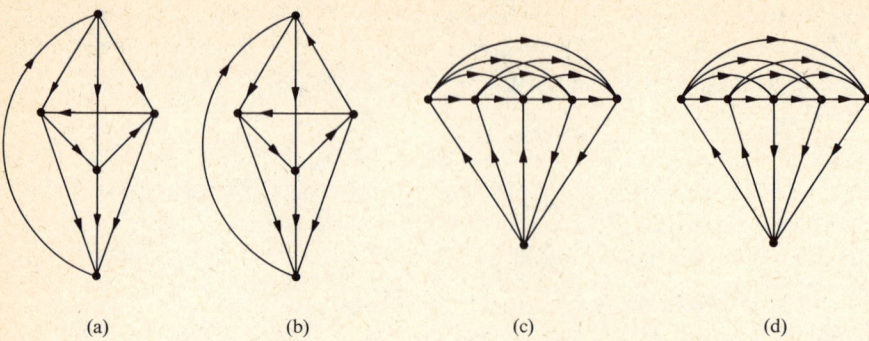

| (a) | (b) | (c) | (d) |

Fig. 16.8. Two pairs of nonreconstructable strong tournaments.

Excursion on Reconstruction of Tournaments

The special case of Ulam's Conjecture for tournaments has been partially solved. Just as for graphs, each tournament T with p points determines p subtournaments $T_i = T - v_i$. We proved* that any nonstrong tournament with at least five points can be reconstructed. However, the conjecture does not hold for strong tournaments with $p = 5$ and 6. This was established by L. W. Beineke and E. M. Parker, who found that the two pairs of tournaments, Fig. 16.8(a, b) and Fig. 16.8(c, d), are counterexamples.

No larger such examples are yet known, and we conjecture that there are none!

EXERCISES

16.1 A digraph is strictly weak if it is weak but not unilateral; it is strictly unilateral if it is unilateral but not strong. Let C_0 contain all disconnected digraphs, C_1 the strictly weak ones, C_2 strictly unilateral, and C_3 those which are strong. Then the maximum and minimum possible number q of arcs among all p point digraphs in connectedness category C_i, $i = 0$ to 3 is given in the following table:

Category	Minimum Number of Arcs	Maximum Number of Arcs
0	0	$(p - 1)(p - 2)$
1	$p - 1$	$(p - 1)(p - 2)$
2	$p - 1$	$(p - 1)^2$
3	p	$p(p - 1)$

(Cartwright and Harary [CH1]

---

* F. Harary and E. M. Palmer, On the problem of reconstructing a tournament from subtournaments, *Monatshefte für Math.* **71**, 14–23 (1967).

16.2 The cartesian product $D_1 \times D_2$ of two digraphs has $V_1 \times V_2$ as its point set, and (u_1, u_2) is adjacent to (v_1, v_2) whenever $[u_1 = v_1$ and u_2 adj $v_2]$ or $[u_2 = v_2$ and u_1 adj $v_1]$. (This is defined just as for graphs in Chapter 2, except that adjacency is directed.) When D is in connectedness category C_n, we write $c(D) = n$. Then $c(D_1 \times D_2) = \min \{c(D_1), c(D_2)\}$ unless $c(D_1) = c(D_2) = 2$ in which case $c(D_1 \times D_2) = 1$.

(Harary and Trauth [HT1])

16.3 No strictly weak digraph contains a point whose removal results in a strong digraph.   (Harary and Ross [HR2])

*16.4 There exists a digraph with outdegree sequence $(s_1, s_2, \cdots, s_p)$, where $p - 1 \geq s_1 \geq s_2 \geq \cdots \geq s_p$, and indegree sequence $(t_1, t_2, \cdots, t_p)$ where every $t_j \leq p - 1$ if and only if $\Sigma s_i = \Sigma t_i$, and for each integer $k < p$,

$$\sum_{i=1}^{k} s_i \leq \sum_{i=1}^{k} \min \{k - 1, t_i\} + \sum_{i=k+1}^{p} \min \{k, t_i\}.$$

(Ryser [R21], Fulkerson [F12])

*16.5 There exists a strong digraph with outdegree and indegree sequences as in the preceding exercise if and only if $\Sigma s_i = \Sigma t_i$, each $s_i > 0$, each $t_i > 0$, and for each integer $k < p$, the following strict inequality holds:

$$\sum_{i=1}^{k} s_i < \sum_{i=1}^{k} t_i + \sum_{i=k+1}^{p} \min \{k, t_i\}.$$

(Beineke and Harary [BH1])

16.6 The *line digraph* $L(D)$ has the arcs of the given digraph D as its points, and x is adjacent to y in $L(D)$ whenever arcs x, y induce a walk in D. Calculate the number of points and arcs of $L(D)$ in terms of D.   (Harary and Norman [HN4])

16.7 The line digraph $L(D)$ of a weak digraph D is isomorphic to D if and only if D or D' is functional.   (Harary and Norman [HN4])

16.8 If D is disconnected, the assertion in the preceding exercise does not hold.

*16.9 Let S and T be disjoint sets of points of D and let $X(S, T)$ be the set of all arcs from S to T. Then D is a line digraph if and only if there are no two-point sets S and T such that $|X(S, T)| = 3$.   (Geller and Harary [GH1], Heuchenne [H42])

16.10 The number of eulerian trails of a digraph D equals the number of hamiltonian cycles of $L(D)$.   (Kasteleyn [K3])

16.11 Let T_1 consist of one point with 2 directed loops. Let $T_2 = L(T_1)$ be the line digraph (more precisely pseudodigraph) of T_1 defined as expected, and recursively let $T_n = L(T_{n-1})$. The structures T_n have been called "teleprinter diagrams." Then the number of eulerian trails in T_n is

$$2^{2^{n-1}-n}.$$
(deBruijn and Ehrenfest [BE1])

*16.12 Every digraph in which id v, od $v \geq p/2$ for all points v is hamiltonian.

(Ghouila-Houri [G7])

16.13 Consider those digraphs in which for every point u, the sum $\Sigma d(u, v)$ of the distances from u is constant. Construct such a digraph which is not point-symmetric.

(Harary [H20])

16.14 The complement $\bar{D}$ and the converse D' both have the same group as D.

16.15 Let A be the adjacency matrix of the line digraph of a complete symmetric digraph. Then $A^2 + A$ has all entries 1.   (Hoffman [H45])

16.16 Two digraphs are *cospectral* if their adjacency matrices have the same characteristic polynomial. There exist just three different cospectral strong digraphs with 4 points.

(F. Harary, C. King, and R. C. Read)

16.17 The *conjunction* $D = D_1 \wedge D_2$ of two digraphs D_1 and D_2 has $V = V_1 \times V_2$ as its point set, and $u = (u_1, u_2)$ is adjacent to $v = (v_1, v_2)$ in D whenever u_1 adj v_1 in D_1 *and* u_2 adj v_2 in D_2. The adjacency matrix A of the conjunction $D = D_1 \wedge D_2$ is the tensor product of the adjacency matrices of D_1 and D_2.

(Harary and Trauth [HT1])

16.18 Let D_1 and D_2 be digraphs and let d_i be the greatest common divisor of the lengths of all the cycles in D_i, $i = 1, 2$. Then the conjunction $D_1 \wedge D_2$ is strong if and only if D_1 and D_2 are strong and d_1 and d_2 are relatively prime.   (McAndrew [M7])

16.19 A digraph is called *primitive* if some power of its adjacency matrix A has all its entries positive. A digraph is primitive if and only if it is strong and the lengths of its cycles have greatest common divisor 1. (see Dulmage and Mendelsohn [DM3, p. 204])

*16.20 Let D be a primitive digraph.

a) If n is the smallest integer such that $A^n > 0$, then $n \le (p - 1)^2 + 1$.

(Wielandt [W17])

b) If n has the maximum possible value $(p - 1)^2 + 1$, then there exists a permutation matrix P such that PAP^{-1} has the form $[a_{ij}]$ where $a_{ij} = 1$ whenever $j = i + 1$ and $a_{p,1} = 1$, but $a_{ij} = 0$ otherwise.

(Dulmage and Mendelsohn [DM3, p. 209])

16.21 An *orientation* of a graph G is an assignment of a direction to each line of G. A graph has a strongly connected orientation if and only if it is connected and bridgeless.

(Robbins [R17])

16.22 Let B be the $p \times q$ incidence matrix of an arbitrary orientation D of a given labeled graph G, so that the entry b_{ij} of B is $+1$ if oriented line x_j is incident to point $v_i - 1$ if x_j is incident from v_i, and 0 otherwise. Then the common cofactor of BB^T is the number of spanning trees of G. (Compare the matrix BB^T with M of Chapter 13.)

(Kirchhoff [K7])

16.23 Recall from Chapter 5 that in a graph G, $\lambda(u, v)$ is the minimum number of lines whose removal separates u and v. Similarly, when u and v are points of a digraph D, let $\vec{\lambda}(u, v)$ be the minimum number of arcs whose removal leaves no path from u to v.

For any orientation D of an eulerian graph G, $\vec{\lambda}(u, v) = \vec{\lambda}(v, u) = \frac{1}{2}\lambda(u, v)$ for every pair of points.   (Nash-Williams [N1])

16.24 Every orientation of an n-chromatic graph G contains a path of length $n - 1$.

(Gallai [G4])

16.25 The scores s_i of a tournament satisfy $\Sigma s_i^2 = \Sigma (p - 1 - s_i)^2$.

16.26 All but two tournaments have a spanning path $v_1 v_2 \cdots v_p$ with a shortcut, the arc $v_1 v_p$. The two exceptions are the cyclic triple and tournament of Fig. 16.8(a).

(B. Grünbaum)

16.27 a) The number of cycles of length 4 in any p point tournament is equal to the number of strong subtournaments with 4 points.

   b) The maximum number of strong subtournaments with 4 points in any p point tournament is $t(p, 4) = \frac{1}{2}(p - 3)t(p, 3)$. See Corollary 16.13(a).

(Beineke and Harary [BH4])

16.28 A group is isomorphic to the point-group of some tournament if and only if it has odd order.

(Moon [M14])

16.29 Let Γ be the point-group and Γ_1 the arc-group of a tournament T. Then Γ_1 is transitive if and only if the pair-group of Γ is transitive.   (Jean [J1])

16.30 Let $t(x)$ and $s(x)$ be the generating functions for tournaments and strong tournaments, respectively. Then

$$s(x) = \frac{t(x)}{1 + t(x)}.$$

(Moon [M16, p. 88])

16.31 Consider a sequence of nonnegative integers $s_1 \leq s_2 \leq \cdots \leq s_p$.

   a) This is the score sequence of some tournament T if and only if

$$\sum_1^p s_i = p(p - 1)/2 \quad \text{and for all } k < p, \quad \sum_1^k s_i \geq k(k - 1)/2.$$

(Landau [L1])

   b) Further, T is strong if and only if for all $k < p$,

$$\sum_1^k s_i > k(k - 1)/2.$$

(Harary and Moser [HM2])

GRAPH DIAGRAMS

> One picture is worth more
> than ten thousand words.
> ANONYMOUS

It is very useful to have diagrams of graphs available for the accumulation of data leading to conjectures. Graphs with fewer than 6 points are easily drawn. The diagrams of 6 point graphs which are presented here were produced by D. W. Crowe, who also was apparently the first to draw all 7 point graphs. In listing the diagrams, no attempt was made to settle the problem of assigning a canonical ordering to the various graphs with p points and q lines. However an index n is assigned to each graph G, with the same index going to the complementary graph $\bar{G}$. Thus the graph $G_{p,q,n}$ is the nth (p, q) graph, and is identified to the right of its diagram by the number n; furthermore $\bar{G}_{p,q,n} = G_{p,\binom{p}{2}-q,n}$. The (4, 3) and (5, 5) graphs are of course exceptions to this rule.

As a supplement to tables of this kind, B. R. Heap developed a program on the computer at the National Physical Laboratory in Middlesex which has produced one card for each graph with 7 points and is in the process of producing graphical cards for $p = 8$. It was found most convenient to code the graphs in adjacency matrix form. The existence of such lists has already proved valuable to investigators using computer methods.

For convenience we present here a table displaying the number of graphs with a given number of points and lines, up through 9 points (cf. Riordan [R15, p. 146]). The entries were obtained using Pólya's formula (15.47).

Table A1

THE NUMBER OF GRAPHS WITH $p \leq 9$ POINTS AND q LINES

q \ p	1	2	3	4	5	6	7	8	9
0	1	1	1	1	1	1	1	1	1
1		1	1	1	1	1	1	1	1
2			1	2	2	2	2	2	2
3			1	3	4	5	5	5	5
4				2	6	9	10	11	11
5				1	6	15	21	24	25
6				1	6	21	41	56	63
7					4	24	65	115	148
8					2	24	97	221	345
9					1	21	131	402	771
10					1	15	148	663	1637
11						9	148	980	3252
12						5	131	1312	5995
13						2	97	1557	10120
14						1	65	1646	15615
15						1	41	1557	21933
16							21	1312	27987
17							10	980	32403
18							5	663	34040
g_p	1	2	4	11	34	156	1044	12346	274668

	$q=0$	$q=1$	$q=2$	$q=3$	$q=4$	$q=5$	$q=6$
$p=1$							
$p=2$							
$p=3$							
$p=4$							

$p = 5$

$q=0$	$q=1$	$q=2$	$q=3$

$q=4$	$q=5$

$p = 5$ (cont.)

	$q = 6$	$q = 7$	$q = 8$

$p = 6$

$q=0$	$q=1$	$q=2$	$q=3$

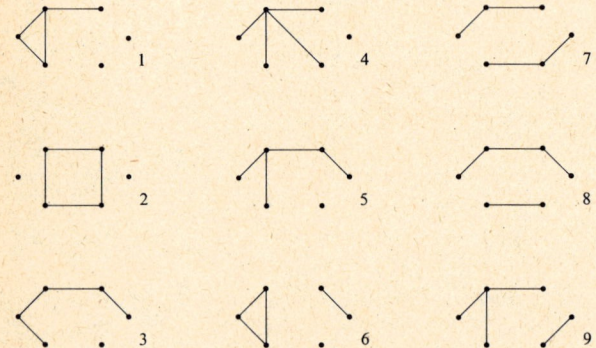

$q=4$

$p=6$ (cont.)

$q=5$

$q=6$

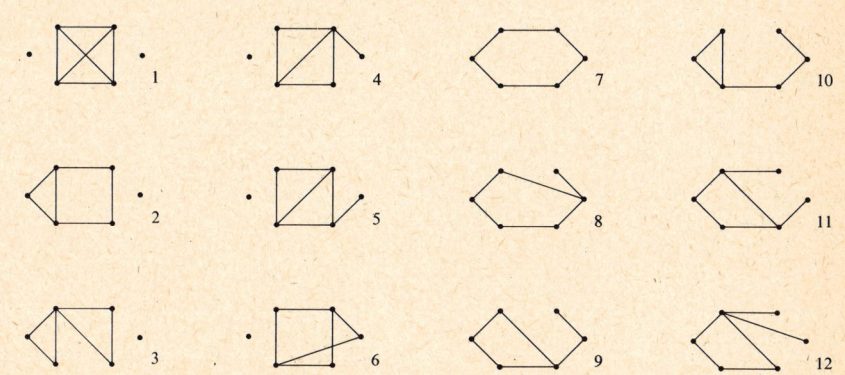

$p=6$ (cont.)

$q=6$

 13  15 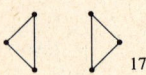 17 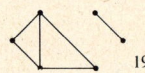 19

 14  16  18 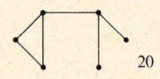 20

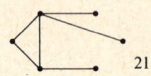 21

$q=7$

 1 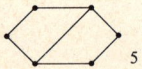 5 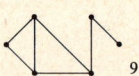 9 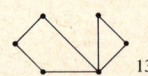 13

 2  6  10  14

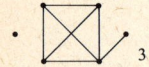 3  7 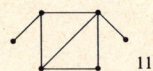 11 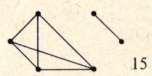 15

 4 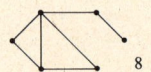 8 12 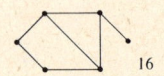 16

$p=6$ (cont.)

$q=7$

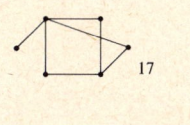 17

 19

 21

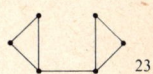 23

 18

 20

 22

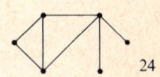 24

$q=8$

 1

 6

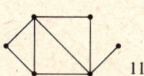 11

 16

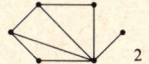 2

 7

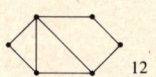 12

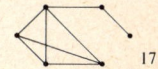 17

 3

 8

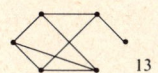 13

 18

 4

 9

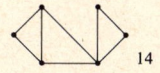 14

 19

 5

 10

 15

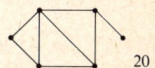 20

$p=6$ (cont.)

$q=8$

 21  22  23 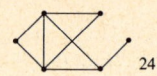 24

$q=9$

 1  7  13  19

 2  8  14  20

 3  9  15  21

 4  10  16

 5  11  17

 6  12  18

$p = 6$ (cont.)

$q = 10$

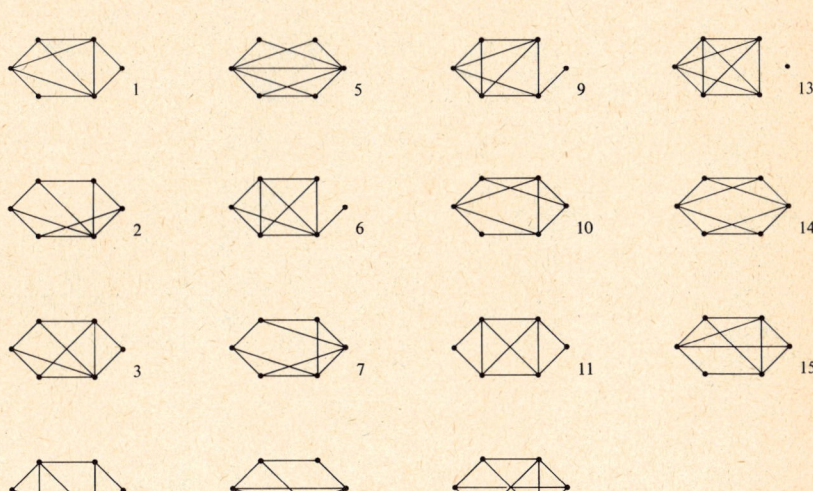

$q = 11$

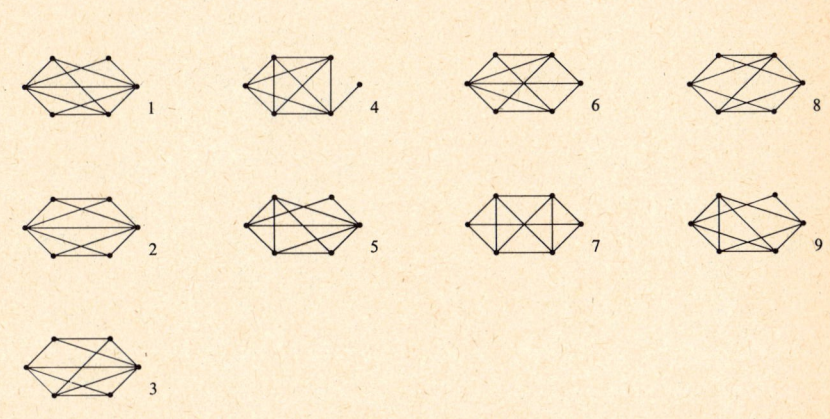

$p = 6$ (cont.)

$q = 12$	$q = 13$	$q = 14$	$q = 15$
1	1		
2	2		
3			
4			
5			

DIGRAPH DIAGRAMS

> The hero jumped on his horse
> and rode off in all directions.
>
> S. Leacock

The digraphs with at most 4 points are listed here according to the number of points and arcs. Indices are assigned to each one in such a way that complements receive the same index, except of course within the (3, 3) and (4, 6) digraphs. The diagrams only go through $p = 4$ because to include those for $p = 5$ would require another book almost the size of the present volume. The following table due to Oberschelp [O1] gives the number of digraphs with p points, $p \leq 8$. The entries may be computed using equations (15.31, 15.32).

Table A2

THE NUMBER
OF DIGRAPHS
WITH $p \leq 8$ POINTS

p	d_p
1	1
2	3
3	16
4	218
5	9 608
6	1 540 944
7	882 033 440
8	1 793 359 192 848

	$q=0$	$q=1$	$q=2$	$q=3$	$q=4$	$q=5$	$q=6$
$p=1$							
$p=2$							
$p=3$							

	$q=0$	$q=1$	$q=2$	$q=3$
$p=4$			1	1 8
			2	2 9
			3	3 10
			4	4 11
			5	5 12
				6 13
				7

$p=4$ (cont.)

$q=4$	$q=5$

$p = 4$ (cont.)

$q = 6$	$q = 7$

$p=4$ (cont.)

$q=7$	$q=8$

$q=9$	$q=10$	$q=11$	$q=12$

TREE DIAGRAMS

> You can't see the forest for the trees.
>
> ANONYMOUS

The diagrams of all the trees with $p \leq 12$ points were developed by Prins and appear as an appendix in his doctoral dissertation [P8]. We present here only those diagrams for $p \leq 10$, which are also given in [HP21]. The ordering of trees with a given number of points is somewhat arbitrary, but in general they are listed by increasing number of points of degree greater than 2. The following table presents the number of trees and rooted trees with p points for $p \leq 26$ (cf. Riordan [R15, p. 138]) and the number of identity trees and homeomorphically irreducible trees for $p \leq 12$ (cf. [HP20]). These numbers were obtained using formulas (15.41), (15.35), (15.51 and 15.52), and (15.47, 15.48, and 15.49) respectively.

Table A3

THE NUMBER OF TREES, ROOTED TREES, IDENTITY TREES,
AND HOMEOMORPHICALLY IRREDUCIBLE TREES WITH p POINTS

p	t_p	T_p	i_p	h_p	p	t_p	T_p
1	1	1	1	1	13	1 301	12 486
2	1	1	0	1	14	3 159	32 973
3	1	2	0	0	15	7 741	87 811
4	2	4	0	1	16	19 320	235 381
5	3	9	0	1	17	48 629	634 847
6	6	20	0	2	18	123 867	1 721 159
7	11	48	1	2	19	317 955	4 688 676
8	23	115	1	4	20	823 065	12 826 228
9	47	286	3	5	21	2 144 505	35 221 832
10	106	719	6	10	22	5 623 756	97 055 181
11	235	1842	15	14	23	14 828 074	268 282 855
12	551	4766	29	26	24	39 299 897	743 724 984
					25	104 636 890	2 067 174 645
					26	279 793 450	5 759 636 510

$p=1$

$p=2$

$p=3$

$p=4$

$p=5$

$p=6$

$p=7$

$p=8$

$p=9$

$p=10$

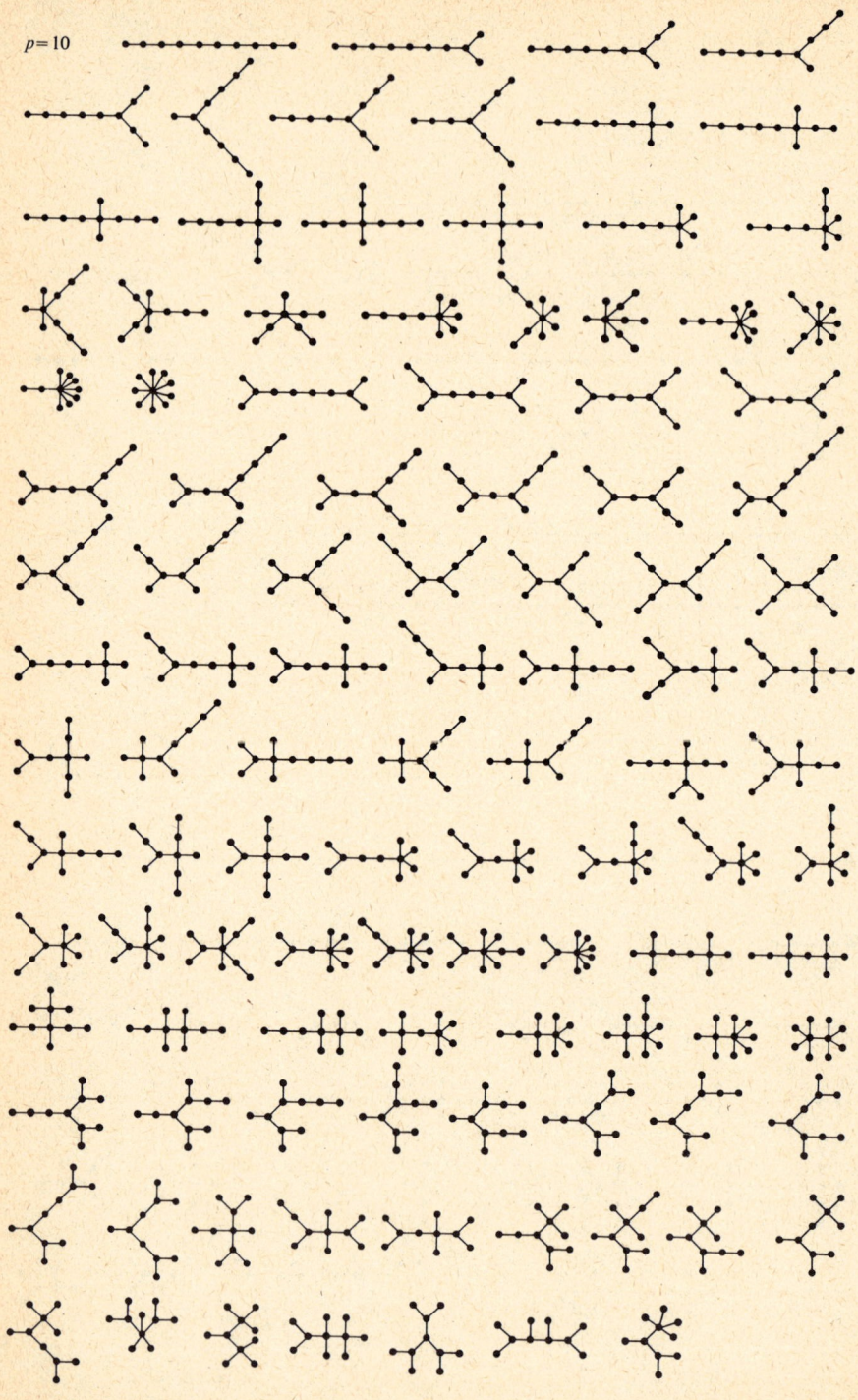

BIBLIOGRAPHY

BIBLIOGRAPHY

And thick and fast
They came at last,
And more and more and more.

L. CARROLL

The references listed below are intended to be those and only those which have been cited in the text. It should be noted, however, that this list is considerably more selective than the exceedingly comprehensive bibliography of graph theory and its applications compiled by Turner [T5]. For the convenience of the reader, each item in this bibliography is followed by one or more numbers in square brackets which indicate the pages where the item is mentioned, following the useful innovation of Grünbaum [G10]. In accordance with the procedure in *Mathematical Reviews*, all books are starred.†

Anderson, S. S.
  [AH1]   (with F. Harary), Trees and unicyclic graphs. *Math. Teacher* **60** (1967), 345–348. [42]

Bäbler, F.
  [B1]   Über eine spezielle Klasse Euler'scher Graphen. *Comment. Math. Helv.* **27** (1953), 81–100. [69]

Balaban, A. T.
  [B2]   Valence-isomerism of cyclopolyenes. *Rev. Roumaine Chim.* **11** (1966), 1097–1116. [62]

Ball, W. W. R.
* [BC1]   (with H. S. M. Coxeter), *Mathematical Recreations and Essays.* Macmillan, New York, 1947. [4]

Barnette, D.
  [B3]   Trees in polyhedral graphs. *Canad. J. Math.* **18** (1966), 731–736. [124] See also p. 68.

---

† One publisher advertised a trigonometry text saying, "This book was starred in Math. Reviews."

Battle, J.

[BHK1]   (with F. Harary and Y. Kodama), Every planar graph with nine points has a nonplanar complement. *Bull. Amer. Math. Soc.* **68** (1962), 569–571.   [108]

[BHKY1]   (with F. Harary, Y. Kodama, J. W. T. Youngs), Additivity of the genus of a graph. *Bull. Amer. Math. Soc.* **68** (1962), 565–568.   [119]

Behzad, M.

[B4]   A criterion for the planarity of a total graph. *Proc. Cambridge Philos. Soc.* **63** (1967), 679–681.   [71, 82, 124]

[BC2]   (with G. Chartrand), Total graphs and traversability. *Proc. Edinburgh Math. Soc.* **15** (1966), 117–120.   [83]

[BC3]   (with G. Chartrand), No graph is perfect. *Amer. Math. Monthly* **74** (1967), 962–963.   [62]

[BCC1]   (with G. Chartrand and J. Cooper), The colour numbers of complete graphs. *J. London Math. Soc.* **42** (1967), 226–228.   [149]

[BCN1]   (with G. Chartrand and E. A. Nordhaus), Triangles in line-graphs and total graphs. *Indian J. Math.* (to appear).   [83]

[BR1]   (with H. Radjavi), The line analog of Ramsey numbers. *Israel J. Math.* **5** (1967), 93–96.   [82]

Beineke, L. W.

[B5]   Decompositions of complete graphs into forests. *Magyar Tud. Akad. Mat. Kutato Int. Kozl.* **9** (1964), 589–594.   [91]

[B6]   The decomposition of complete graphs into planar subgraphs. Chapter 4 in *Graph Theory and Theoretical Physics* (F. Harary, ed.) Academic Press, London, 1967, pp. 139–154.   [120]

[B7]   Complete bipartite graphs: decomposition into planar subgraphs. Chapter 7 in *A Seminar in Graph Theory* (F. Harary, ed.) Holt, Rinehart and Winston, New York, 1967, pp. 42–53.   [120]

[B8]   Derived graphs and digraphs. *Beiträge zur Graphentheorie* (H. Sachs, H. Voss, and H. Walther, eds.) Teubner, Leipzig, 1968, pp. 17–33.
                                                                        [71, 74]

[BG1]   (with R. K. Guy), The coarseness of $K_{m,n}$. *Canad. J. Math.* (to appear).
                                                                    [121, 125]

[BH1]   (with F. Harary), Local restrictions for various classes of directed graphs. *J. London Math. Soc.* **40** (1965), 87–95.   [209]

[BH2]   (with F. Harary), Inequalities involving the genus of a graph and its thickness. *Proc. Glasgow Math. Assoc.* **7** (1965), 19–21.   [118, 125]

[BH3]   (with F. Harary), The genus of the n-cube. *Canad. J. Math.* **17** (1965), 494–496.   [119]

[BH4]   (with F. Harary), The maximum number of strongly connected sub-tournaments. *Canad. Math. Bull.* **8** (1965), 491–498.   [207, 211]

[BH5]   (with F. Harary), The thickness of the complete graph. *Canad. J. Math.* **17** (1965), 850–859.   [120]

[BH6]   (with F. Harary), The connectivity function of a graph. *Mathematika* **14** (1967), 197–202.   [45, 51]

[BHM1]   (with F. Harary and J. W. Moon), On the thickness of the complete bipartite graph. *Proc. Cambridge Philos. Soc.* **60** (1964), 1–5.   [120]

[BHP1]   (with F. Harary and M. D. Plummer), On the critical lines of a graph. *Pacific J. Math.* **22** (1967), 205–212.   [98]
[BM1]   (with J. W. Moon), The number of labelled *k*-trees. *Proof Techniques in Graph Theory* (F. Harary, ed.) Academic Press, New York, 1969.
                                                                          [195]
[BP1]   (with R. E. Pippert), The enumeration of labelled *k*-dimensional trees. *J. Combinatorial Theory* (to appear).   [195]
[BP2]   (with M. D. Plummer), On the 1-factors of a nonseparable graph. *J. Combinatorial Theory* **2** (1967), 285–289.   [85, 92]
          See also [GB1].

Benzer, S.
  [B9]   On the topology of the genetic fine structure. *Proc. Nat. Acad. Sci. USA* **45** (1959), 1607–1620.   [20]

Berge, C.
  [B10]   Two theorems in graph theory. *Proc. Nat. Acad. Sci. USA* **43** (1957), 842–844.   [96]
* [B11]   *Théorie des Graphes et ses Applications.* Dunod, Paris, 1958.   [44]
* [B12]   *The Theory of Graphs and its Applications.* Methuen, London, 1962.
                                                            [21, 97, 100, 128, 201]
* [BG2]   (with A. Ghouila-Houri), *Programming, Games, and Transportation Networks.* Methuen, London, 1965.   [7]

Binet, J. P. M. See p. 153.

Birkhoff, G.
* [B13]   *Lattice Theory.* Amer. Math. Soc. Colloq. Publ., Vol. 25, Third Edition, Providence, 1967.   [54]

Birkhoff, G. D.
  [BL1]   (with D. Lewis), Chromatic polynomials. *Trans. Amer. Math. Soc.* **60** (1946), 355–451.   [145]

Boland, J.
  [BL2]   (with C. Lekkerkerker), Representation of a finite graph by a set of intervals on the real line. *Fund. Math.* **51** (1962), 45–64.   [21]

Bollobás, B.
  [B14]   On graphs with at most three independent paths connecting any two vertices. *Studia Sci. Math. Hungar.* **1** (1966), 137–140.   [55, 56]

Bondy, J. A.
  [B15]   On Kelly's congruence theorem for trees. *Proc. Cambridge Philos. Soc.* **65** (1969), 1–11.   [41]
          See also p. 149.

Bosak, J. See p. 68.

Bott, R.
  [BM2]   (with J. P. Mayberry), Matrices and trees. *Economic activity analysis.* (O. Morgenstern, ed.) New York, Wiley, 391–400.   [203]

Brooks, R. L.
  [B16]   On colouring the nodes of a network. *Proc. Cambridge Philos. Soc.* **37** (1941) 194–197.   [128]

[BSST1]   (with C. A. B. Smith, A. H. Stone, and W. T. Tutte), The dissection of rectangles into squares. *Duke Math. J.* **7** (1940), 312–340.   [158]
See also p. 123.

Brualdi, R. A.
[B17]   Kronecker products of fully indecomposable matrices and of ultrastrong digraphs, *J. Combinatorial Theory* **2** (1967), 135–139.   [21]

deBruijn, N. G.
[B18]   Generalization of Pólya's fundamental theorem in enumeration combinatorial analysis. *Indagationes Math.* **21** (1959), 59–69.   [191]
[B19]   Pólya's theory of counting. *Applied Combinatorial Mathematics* (E. F. Beckenbach, ed.) Wiley, New York, 1964, pp. 144–184.   [191]
[BE1]   (with T. van Aardenne-Ehrenfest), Circuits and trees in oriented graphs. *Simon Stevin* **28** (1951), 203–217.   [195, 204, 209]

Burnside, W.
* [B20]   *Theory of Groups of Finite Order* (second edition) Cambridge Univ. Press, Cambridge, 1911.   [180]

Cameron, J. See p. 25.

Camion, P.
[C1]   Chemie et circuits hamiltoniens des graphes complets. *C.R. Acad. Sci. Paris* **249** (1959), 2151–2152.   [206]

Carlitz, L.
[CR1]   (with J. Riordan), The number of labelled two-terminal series-parallel networks. *Duke Math. J.* **23** (1955), 435–445.   [195]

Cartwright D.
[CH1]   (with F. Harary), The number of lines in a digraph of each connectedness category. *SIAM Review* **3** (1961), 309–314.   [208]
[CH2]   (with F. Harary), On colorings of signed graphs. *Elem. Math.* **23** (1968), 85–89.   [138, 149]
See also [HNC1].

Cauchy, A. L. See p. 153.

Cayley, A.
[C2]   On the theory of the analytical forms called trees. *Philos. Mag.* **13** (1857), 19–30. *Mathematical Papers*, Cambridge **3** (1891), 242–246.   [3, 187]
[C3]   On the mathematical theory of isomers. *Philos. Mag.* **67** (1874), 444–446. *Mathematical Papers*, Cambridge **9** (1895), 202–204; and **9** (1895), 427–460.   [3]
[C4]   The theory of groups, graphical representation. *Mathematical Papers*, Cambridge **10** (1895), 26–28.   [168]
[C5]   On the analytical forms called trees. *Amer. Math. J.* **4** (1881), 266–268. *Mathematical Papers*, Cambridge **11** (1896), 365–367.   [188]
[C6]   A theorem on trees. *Quart. J. Math.* **23** (1889), 376–378. *Mathematical Papers*, Cambridge **13** (1897), 26–28.   [179, 195]

Chang, L. C.
[C7]   The uniqueness and nonuniqueness of the triangular association scheme. *Sci. Record* **3** (1959), 604–613.   [78]

Chartrand, G.
  [C8]   A graph-theoretic approach to a communications problem. *J. SIAM Appl. Math.* **14** (1966), 778–781.   [44]
  [C9]   On Hamiltonian line graphs. *Trans. Amer. Math. Soc.* (to appear).
                                                                        [81]
  [CG1]   (with D. Geller), Uniquely colorable planar graphs. *J. Combinatorial Theory* (to appear).   [138, 140, 149]
  [CGH1] (with D. Geller and S. Hedetniemi), A generalization of the chromatic number. *Proc. Cambridge Philos. Soc.* **64** (1968), 265–271.   [149]
  [CGH2] (with D. Geller and S. Hedetniemi), Graphs with forbidden subgraphs. *J. Combinatorial Theory* (to appear).   [25, 100, 124]
  [CH3]   (with F. Harary), Planar permutation graphs. *Ann. Inst. Henri Poincaré* Sec. B **3** (1967), 433–438.   [107, 175, 177]
  [CH4]   (with F. Harary), Graphs with prescribed connectivities. *Theory of Graphs* (P. Erdös and G. Katona, eds.) Akadémiai Kiadó, Budapest, 1968, 61–63.   [44]
  [CKK1] (with S. F. Kapoor and H. V. Kronk), The Hamiltonian hierarchy. (to appear).   [70]
  [CKL1] (with A. Kaugars and D. R. Lick), Critically *n*-connected graphs. *Proc. Amer. Math. Soc.* (submitted).   [56]
  [CK1]   (with H. V. Kronk), Randomly traceable graphs. *J. SIAM Appl. Math.* (to appear).   [70]
  [CS1]   (with M. J. Stewart), The connectivity of line-graphs. *Math. Ann.* (submitted).   [83]
          See also [BC2], [BC3], [BCC1], [BCN1], [KC1], and p. 77.

Cooper, J. See [BCC1].

Courant, R.
* [CR1]   (with H. E. Robbins) *What is Mathematics?* Oxford U. Press, London, 1941.   [117]

Coxeter, H. S. M.
  [C10]   Self-dual configurations and regular graphs. *Bull. Amer. Math. Soc.* **56** (1950), 413–455.   [175]
          See also [BC1].

Crowe, D. W. See p. 213.

Danzer, L.
  [DK1]   (with V. Klee), Lengths of snakes in boxes. *J. Combinatorial Theory* **2** (1967), 258–265.   [25]

Dauber, E. See p. 172.

Davis, R. L.
  [D1]   The number of structures of finite relations. *Proc. Amer. Math. Soc.* **4** (1953), 486–495.   [195]
  [D2]   Structures of dominance relations. *Bull. Math. Biophys.* **16** (1954), 131–140.   [195]

Descartes, B.

[D3]   Solution to advanced problem no. 4526. *Amer. Math. Monthly* **61** (1954), 352.   [128]

Dilworth, R. P.

[D4]   A decomposition theorem for partially ordered sets. *Ann. Math.* **51** (1950), 161–166.   [54]

Dirac, G. A.

[D5]   A property of 4-chromatic graphs and some remarks on critical graphs. *J. London Math. Soc.* **27** (1952), 85–92.   [135, 141]

[D6]   Some theorems on abstract graphs. *Proc. London Math. Soc.*, Ser. 3, **2** (1952), 69–81.   [68, 142]

[D7]   The structure of k-chromatic graphs. *Fund. Math.* **40** (1953), 42–55.   [128, 142]

[D8]   In abstrakten Graphen verhandene vollständige 4-Graphen und ihre Unterteilungen. *Math. Nachr.* **22** (1960), 61–85.   [45, 149]

[D9]   Généralisations du théorème de Menger. *C.R. Acad. Sci. Paris* **250** (1960), 4252–4253.   [50]

[D10]   Short proof of Menger's graph theorem. *Mathematika* **13** (1966), 42–44.   [47]

[D11]   On the structure of k-chromatic graphs. *Proc. Cambridge Philos. Soc.* **63** (1967), 683–691.   [149]

[DS1]   (with S. Schuster), A theorem of Kuratowski. *Nederl. Akad. Wetensch. Proc.* Ser. A **57** (1954), 343–348.   [108]

Dulmage, A. L.

[DM1]   (with N. S. Mendelsohn), Coverings of bipartite graphs. *Canad. J. Math.* **10** (1958), 517–534.   [98, 99]

[DM2]   (with N. S. Mendelsohn), On the inversion of sparse matrices. *Math. Comp.* **16** (1962), 494–496.   [205]

[DM3]   (with N. S. Mendelsohn), Graphs and matrices. Chapter 6 in *Graph Theory and Theoretical Physics* (F. Harary, ed.) Academic Press, London, 1967, pp. 167–229.   [210]

Edmonds, J.

[E1]   Existence of k-edge connected ordinary graphs with prescribed degrees. *J. Res. Nat. Bur. Stand.*, Sect.B **68** (1964), 73–74.   [63]

Ehrenfest, T. van Aardenne-. See [BE1].

Elias, P.

[EFS1]   (with A. Feinstein and C. E. Shannon), A note on the maximum flow through a network. *IRE Trans. Inform. Theory*, IT-2, (1956), 117–119.   [49]

Erdös, P.

[E2]   Graph theory and probability II. *Canad. J. Math.* **13** (1961), 346–352.   [128]

[E3]   Extremal problems in graph theory. Chapter 8 of *A Seminar on Graph*

Theory (F. Harary, ed.) Holt, Rinehart and Winston, New York, 1967, pp. 54–59. [17, 18, 25]

[E4]   Applications of probabilistic methods to graph theory. Chapter 9 in *A Seminar on Graph Theory* (F. Harary, ed.) Holt, Rinehart and Winston, New York, 1967, pp. 60–64. [24]

[EG1]   (with T. Gallai), Graphs with prescribed degrees of vertices (Hungarian). *Mat. Lapok* **11** (1960), 264–274. [58, 59]

[EGP1]   (with A. Goodman and L. Pósa), The representation of a graph by set intersections. *Canad. J. Math.* **18** (1966), 106–112. [20]

[EH1]   (with A. Hajnal), On chromatic numbers of graphs and set-systems. *Acta Math. Acad. Sci. Hungar.* **17** (1966), 61–99. [148]

[ER1]   (with A. Rényi), Asymmetric graphs. *Acta Math. Acad. Sci. Hungar.* **14** (1963), 295–315. [17]

[ES1]   (with G. Szekeres), A combinatorial problem in geometry. *Compositio Math.* **2** (1935), 463–470. [16]
See also p. 121.

Euler, L.
[E5]   Solutio problematis ad geometriam situs pertinentis. *Comment. Academiae Sci. I. Petropolitanae* **8** (1736), 128–140. *Opera Omnia* Series I-7 (1766), 1–10. [2]

[E6]   The Königsberg bridges. *Sci. Amer.* **189** (1953), 66–70. [14]
See also p. 64 and 103.

Evans, J. W.
[EHL1]   (with F. Harary and M. S. Lynn), On the computer enumeration of finite topologies. *Comm. Assoc. Comp. Mach.* **10** (1967), 295–298. [195]

Fagen, R. E.   See [AFPR1].

Fáry, I.
[F1]   On straight line representation of planar graphs. *Acta Sci. Math. Szeged.* **11** (1948), 229–233. [106]

Feinstein, A.   See [EFS1].

Feller, W.
* [F2]   *An Introduction to Probability Theory and its Applications*, Vol. 1, (2nd ed.) Wiley, New York, 1957. [6]

Feynmann, R. P.
[F3]   Space-time approaches to quantum electrodynamics. *Physical Review* **76** (1949), 769–789. [6]

Finck, H. J.
[F4]   Über die chromatischen Zahlen eines Graphen und Seines Komplements, I and II. *Wiss. Z. T. H. Ilmenau* **12** (1966), 243–251. [148]

[FS1]   (with H. Sachs), Über eine von H. S. Wilf angegebene Schranke für die chromatische Zahl endlicher Graphen. *Math. Nachr.* **39** (1969), 373–386. [132]

Folkman, J.
[F5]   Regular line-symmetric graphs. *J. Combinatorial Theory* (to appear). [173]

Ford, G. W.
  [FNU1]   (with R. Z. Norman and G. E. Uhlenbeck), Combinatorial problems in the theory of graphs, II. *Proc. Nat. Acad. Sci. USA* **42** (1956), 203–208.
        [195]

Ford, L. R.
  [FF1]   (with D. R. Fulkerson), Maximal flow through a network, *Canad. J. Math.* **8** (1956), 399–404.   [49]
  * [FF2]   (with D. R. Fulkerson), *Flows in Networks.* Princeton University Press, Princeton, 1962.   [7, 52]

Foster, R. M.
  [F6]   Geometrical circuits of electrical networks. *Trans. Amer. Inst. Elec. Engrs.* **51** (1932), 309–317.   [171]

Foulkes, J. D.
  [F7]   Directed graphs and assembly schedules. *Proc. Symp. Appl. Math.,* Amer. Math. Soc. **10** (1960), 281–289.   [206]

Frucht, R.
  [F8]   Herstellung von Graphen mit vorgegebener abstrakten Gruppe. *Compositio Math.* **6** (1938), 239–250.   [168]
  [F9]   Graphs of degree three with a given abstract group. *Canad. J. Math.* **1** (1949), 365–378.   [170]
  [F10]   On the groups of repeated graphs. *Bull. Amer. Math. Soc.* **55** (1949), 418–420.   [165]
  [F11]   A one-regular graph of degree three. *Canad. J. Math.* **4** (1952), 240–247.
        [175]
  [FH1]   (with F. Harary), On the corona of two graphs. *Aequationes Math.* (to appear).   [167]
        See also p. 168.

Fulkerson, D. R.
  [F12]   Zero-one matrices with zero trace. *Pacific J. Math.* **10** (1960), 831–836.
        [209]
  [F13]   Networks, frames, and blocking systems. *Mathematics of the Decision Sciences.* Vol. II, Lectures in Applied Mathematics, (G. B. Dantzig and A. F. Scott, eds.) 303–334.   [52]
        See also [FF1], and [FF2].

Gaddum, J. W. See [NG1].

Gallai, T.
  [G1]   On factorisation of graphs. *Acta Math. Acad. Sci. Hungar.* **1** (1950), 133–153.   [85]
  [G2]   Über extreme Punkt- und Kantenmengen. *Ann. Univ. Sci. Budapest,* Eötvös Sect. Math. **2** (1959), 133–138.   [95]
  [G3]   Elementare relationen bezüglich der glieder und trennenden punkte von graphen. *Magyar Tud. Akad. Mat. Kutato Int. Kozl.* **9** (1964), 235–236.
        [31, 36]
  [G4]   On directed paths and circuits. *Theory of Graphs* (P. Erdös and G. Katona, eds.) Akadémiai Kiadó, Budapest, 1968. Also Academic Press, New York, 1968, pp. 115–119.   [149, 210]

[GM1]   (with A. N. Milgram), Verallgemeinerung eines graphen theoretischen
          Satzes von Rédei, *Acta Scient. Math.* **21** (1960), 181–186.   [206]
          See also [EG1].

Gaudin, T.
[GHR1]   (with J. C. Herz and P. Rossi), Solution du problème no. 29. *Rev. Franc.
          Rech. Oper.* **8** (1964), 214–218.   [70]

Geller, D. P.
[G5]   Outerplanar graphs. (to appear).   [108]
[GH1]   (with F. Harary), Arrow diagrams are line digraphs. *J. SIAM Appl.
          Math.* **16** (1968), 1141–1145.   [209]
          See also [CG1], [CGH1], [CGH2], and p. 83.

Gerencsér, L.
[G6]   On coloring problems (Hungarian). *Mat. Lapok* **16** (1965), 274–277.
                                                                          [101]

Gewirtz, A.
[GQ1]   (with L. V. Quintas), Connected extremal edge graphs having symmetric
          automorphism group. *Recent Progress in Combinatorics* (W. T. Tutte,
          ed.) Academic Press, New York, 1969.   [176]

Ghouila-Houri, A.
[G7]   Une condition suffisante d'éxistence d'un circuit hamiltonien. *C. R.
          Acad. Sci. Paris* **251** (1960), 495–497.   [209]
          See also [BG2].

Gilbert, E. N.
[GR1]   (with J. Riordan), Symmetry types of periodic sequences. *Illinois J. Math.*
          5 (1961), 657–665.   [195]

Gilmore, P.
[GH2]   (with A. J. Hoffman), A characterization of comparability graphs and
          interval graphs. *Canad. J. Math.* **16** (1964), 539–548.   [21]

Goodman, A. See [EGP1].

Graver, J. E.
[GY1]   (with J. Yackel), Some graph theoretic results associated with Ramsey's
          theorem. *J. Combinatorial Theory* **4** (1968), 125–175.   [16]

Grossman, J. W. See [HG1].

Grötzsch, H.
[G8]   Ein Dreifarbensatz für dreikreisfreie Netze auf der Kugel. *Wiss. Z.
          Martin-Luther Univ. Halle-Wittenberg. Math. Naturwiss. Reihe* **8** (1958),
          109–120.   [131]

Grünbaum, B.
*[G9]   Grötzsch's theorem on 3-colorings. *Michigan Math. J.* **10** (1963),
          303–310.   [131]
[G10]   *Convex Polytopes*, Wiley (Interscience), New York, 1967.
                                                                  [68, 117, 235]
          See also p. 211.

Gupta, R. P.
  [G11]   Independence and covering numbers of line graphs and total graphs. *Proof Techniques in Graph Theory* (F. Harary, ed.) Academic Press, New York, 1969.   [97]

Guy, R. K.
* [G12]   The decline and fall of Zarankiewicz's theorem. *Proof Techniques in Graph Theory* (F. Harary, ed.) Academic Press, New York, 1969.   [122]
  [GB1]   (with L. W. Beineke), The coarseness of the complete graph. *Canad. J. Math.* **20** (1966), 888–894.   [121]
         See also [BG1].

Hadwiger, H.
  [H1]   Über eine Klassifikation der Streckenkomplexe. *Vierteljschr. Naturforsch. Ges. Zürich* **88** (1943), 133–142.   [135]

Hajnal, A. See [EH1].

Hajós, G.
  [H2]   Über eine Art von Graphen. *Internat. Math. Nachr.* **2** (1957), 65.   [20]
  [H3]   Über eine Konstruktion nicht n-färbbarer Graphen. *Wiss. Z. Martin Luther Univ. Halle-Wittenberg Math.-Natur. Reihe.* **10** (1961), 116–117.   [143]

Hakimi, S.
  [H4]   On the realizability of a set of integers as degrees of the vertices of a graph. *J. SIAM Appl. Math.* **10** (1962), 496–506.   [58, 62, 63]

Halin, R.
  [H5]   A theorem on n-connected graphs. *J. Combinatorial Theory* (to appear).   [56]

Hall, D. W.
  [H6]   A note on primitive skew curves. *Bull. Amer. Math. Soc.* **49** (1943), 935–937.   [123]

Hall, M.
* [H7]   *Combinatorial Theory*, Blaisdell, Waltham, 1967.   [16, 56]

Hall, P.
  [H8]   On representations of subsets. *J. London Math. Soc.* **10** (1935), 26–30.   [53]

Hamelink, R. C.
  [H9]   A partial characterization of clique graphs. *J. Combinatorial Theory* (to appear).   [20]

Harary, F.
  [H10]   On the notion of balance of a signed graph. *Mich. Math. J.* **2** (1953), 143–146.   [195]
  [H11]   The number of linear, directed, rooted, and connected graphs. *Trans. Amer. Math. Soc.* **78** (1955), 445–463.   [185, 186, 195]
  [H12]   Note on the Pólya and Otter formulas for enumerating trees. *Mich. Math. J.* **3** (1956), 109–112.   [189]
  [H13]   On the number of dissimilar line-subgraphs of a given graph. *Pacific J. Math.* **6** (1956), 57–64.   [195]

[H14]   The number of dissimilar supergraphs of a linear graph. *Pacific J. Math.*
7 (1957), 903–911.   [195]

[H15]   Structural duality. *Behavioral Sci.* **2** (1957), 255–265.   [30]

[H16]   The number of oriented graphs. *Mich. Math. J.* **4** (1957), 221–224.
                                                                          [195]

[H17]   On arbitrarily traceable graphs and directed graphs. *Scripta Math.* **23**
(1957), 37–41.   [69]

[H18]   On the number of bicolored graphs. *Pacific J. Math.* **8** (1958), 743–755.
                                                                     [163, 177]

[H19]   On the number of dissimilar graphs between a given graph-subgraph
pair. *Canad. J. Math.* **10** (1958), 513–516.   [195]

[H20]   Status and contrastatus. *Sociometry* **22** (1959), 23–43.   [209]

[H21]   On the group of the composition of two graphs. *Duke Math. J.* **26**
(1959), 29–34.   [21, 166]

[H22]   An elementary theorem on graphs. *Amer. Math. Monthly* **66** (1959),
405–407.   [31]

[H23]   The number of functional digraphs. *Math. Ann.* **138** (1959), 203–210.
                                                                          [195]

[H24]   Unsolved problems in the enumeration of graphs. *Publ. Math. Inst.
Hung. Acad. Sci.* **5** (1960), 63–95.   [192]

[H25]   A graph theoretic approach to matrix inversion by partitioning. *Numer.
Math.* **4** (1962), 128–135.   [205]

[H26]   The maximum connectivity of a graph. *Proc. Nat. Acad. Sci. USA* **48**
(1962), 1142–1146.   [44]

[H27]   The determinant of the adjacency matrix of a graph. *SIAM Review* **4**
(1962), 202–210.   [151]

[H28]   A characterization of block-graphs. *Canad. Math. Bull.* **6** (1963), 1–6.
                                                                           [30]

[H29]   On the reconstruction of a graph from a collection of subgraphs. *Theory
of Graphs and its Applications* (M. Fiedler, ed.). Prague, 1964, 47–52;
reprinted, Academic Press, New York, 1964.   [12, 41]

[H30]   Combinatorial problems in graphical enumeration. *Applied Combinatorial Mathematics* (E. F. Beckenbach, ed.) Wiley, New York, 1964,
pp. 185–220.   [192]

[H31]   Applications of Pólya's theorem to permutation groups. Chapter 5 in
A Seminar on Graph Theory (F. Harary, ed.) Holt, Rinehart and Winston,
New York, 1967, pp. 25–33.   [183, 184, 185, 195]

[H32]   Graphical enumeration problems. Chapter 1 in *Graph Theory and
Theoretical Physics* (F. Harary, ed.) Academic Press, London, 1967,
pp. 1–41.   [192, 194]

[H33]   Variations on a theorem by Menger. *J. SIAM Appl. Math.* (to appear).
                                                                           [55]

[HH1]   (with S. Hedetniemi), The achromatic number of a graph. *J. Combinatorial Theory* (to appear).   [128, 144]

[HHP1]   (with S. Hedetniemi and G. Prins), An interpolation theorem for graphical
homomorphisms. *Port. Math.* (to appear).   [143, 144, 149]

[HHR1]   (with S. T. Hedetniemi and R. W. Robinson), Uniquely colorable graphs. *J. Combinatorial Theory* (to appear).   [139, 149]

[HKT1]   (with R. M. Karp and W. T. Tutte), A criterion for planarity of the square of a graph. *J. Combinatorial Theory* **2** (1967), 395–405.   [124]

[HK1]   (with Y. Kodama), On the genus of an *n*-connected graph. *Fund. Math.* **54** (1964), 7–13.   [47, 119]

[HM1]   (with B. Manvel), On the number of cycles in a graph. *Math. časopis* (to appear).   [25, 158]

[HM2]   (with L. Moser), The theory of round robin tournaments. *Amer. Math. Monthly* **73** (1966), 231–246.   [206, 207, 211]

[HMR1]   (with A. Mowshowitz and J. Riordan), Labeled trees with unlabeled endpoint. *J. Combinatorial Theory* **6** (1969), 60–64.   [195]

[HN1]   (with C. St. J. A. Nash-Williams), On Eulerian and Hamiltonian graphs and line-graphs. *Canad. Math. Bull.* **8** (1965), 701–709.   [81, 83]

[HN2]   (with R. Z. Norman), The dissimilarity characteristic of Husimi trees. *Ann. of Math.* **58** (1953), 134–141.   [42, 195]

[HN3]   (with R. Z. Norman), Dissimilarity characteristic theorems for graphs. *Proc. Amer. Math. Soc.* **11** (1960), 332–334.   [197]

[HN4]   (with R. Z. Norman), Some properties of line digraphs. *Rend. Circ. Mat. Palermo* **9** (1961), 161–168.   [209]

* [HNCl]   (with R. Z. Norman and D. Cartwright), *Structural Models: an introduction to the theory of directed graphs.* Wiley, New York, 1965. [6, 198]

[HP1]   (with E. M. Palmer), The number of graphs rooted at an oriented line. *ICC Bull.* **4** (1965), 91–98.   [195]

[HP2]   (with E. M. Palmer), A note on similar points and similar lines of a graph. *Rev. Roum. Math. Pures et Appl.* **10** (1965), 1489–1492.   [176]

[HP3]   (with E. M. Palmer), The smallest graph whose group is cyclic. *Czech. Math. J.* **16** (1966), 70–71.   [170, 176]

[HP4]   (with E. M. Palmer), On the number of orientations of a given graph. *Bull. Acad. Polon. Sci. Ser. Sci. Math. Astronom. Phys.* **14** (1966), 125–128.   [195]

[HP5]   (with E. M. Palmer), On similar points of a graph. *J. Math. Mech.* **15** (1966), 623–630.   [171]

[HP6]   (with E. M. Palmer), The reconstruction of a tree from its maximal proper subtrees. *Canad. J. Math.* **18** (1966), 803–810.   [41]

[HP7]   (with E. M. Palmer), Enumeration of locally restricted digraphs. *Canad. J. Math.* **18** (1966), 853–860.   [195]

[HP8]   (with E. M. Palmer), The power group enumeration theorem. *J. Combinatorial Theory* **1** (1966), 157–173.   [191]

[HP9]   (with E. M. Palmer), Enumeration of self-converse digraphs. *Mathematika* **13** (1966), 151–157.   [195]

[HP10]   (with E. M. Palmer), The groups of the small digraphs. *J. Indian Statist. Assoc.* **4** (1966), 155–169.   [176]

[HP11]   (with E. M. Palmer), The enumeration methods of Redfield. *Amer. J. Math.* **89** (1967), 373–384.   [178]

[HP12]   (with E. M. Palmer), Enumeration of finite automata. *Information and Control* **10** (1967), 499–508.   [192, 195]

[HP13]   (with E. M. Palmer), On the number of balanced signed graphs. *Bull. Math. Biophysics* **29** (1967), 759–765.   [195]

[HP14]   (with E. M. Palmer), On the group of a composite graph. *Studia Sci. Math. Hungar.* **3** (to appear).   [21]

[HP15]   (with E. M. Palmer), On the point-group and line-group of a graph. *Acta Math. Acad. Sci. Hung.* **19** (1968), 263–269.   [162, 177]

[HP16]   (with E. M. Palmer), Note on the number of forests. *Mat. časopis* (to appear).   [195]

[HP17]   (with E. M. Palmer), On acyclic simplicial complexes. *Mathematika* **15** (1968), 119–122.   [195]

[HPR1]   (with E. M. Palmer and R. C. Read), The number of ways to label a structure. *Psychometrika* **32** (1967), 155–156.   [180]

[HP18]   (with M. D. Plummer), On the point-core of a graph. *Math. Z.* **94** (1966), 382–386.   [101]

[HP19]   (with M. D. Plummer), On the core of a graph. *Proc. London Math. Soc.* **17** (1967), 305–314.   [98, 99, 101]

[HP20]   (with G. Prins), The number of homeomorphically irreducible trees, and other species. *Acta Math.* **101** (1959), 141–162.   [189, 231]

[HP21]   (with G. Prins), Enumeration of bicolourable graphs. *Canad. J. Math.* **15** (1963), 237–248.   [195]

[HP22]   (with G. Prins), The block-cutpoint-tree of a graph. *Publ. Math. Debrecen* **13** (1966), 103–107.   [36, 195]

[HPT1]   (with G. Prins and W. T. Tutte), The number of plane trees. *Indagationes Math.* **26** (1964), 319–329.   [195]

[HR1]   (with R. C. Read), The probability of a given 1-choice structure. *Psychometrika* **31** (1966), 271–278.   [180]

[HR2]   (with I. C. Ross), A description of strengthening and weakening group members. *Sociometry* **22** (1959), 139–147.   [209]

[HT1]   (with C. A. Trauth, Jr.), Connectedness of products of two directed graphs. *J. SIAM Appl. Math.* **14** (1966), 250–254.   [21, 209, 210]

[HT2]   (with W. T. Tutte), The number of plane trees with a given partition. *Mathematika* **11** (1964), 99–101.   [195]

[HT3]   (with W. T. Tutte), A dual form of Kuratowski's theorem. *Canad. Math. Bull.* **8** (1965), 17–20, 373.   [113]

[HT4]   (with W. T. Tutte), On the order of the group of a planar map. *J. Combinatorial Theory* **1** (1966), 394–395.   [177]

[HU1]   (with G. E. Uhlenbeck), On the number of Husimi trees, I. *Proc. Nat. Acad. Sci., USA* **39** (1953), 315–322.   [195]

[HW1]   (with G. Wilcox), Boolean operations on graphs. *Math. Scand.* **20** (1967), 41–51.   [21]

           See also [AH1], [BHK1], [BHKY1], [BH1], [BH2], [BH3], [BH4], [BH5], [BH6], [BHM1], [BHP1], [CH1], [CH2], [CH3], [CH4], [EHL1], [FH1], [GH1], and pp. 158 and 208.

Harrison, M. A.
[H34]   A census of finite automata. *Canad. J. Math.* **17** (1965), 100–113.   [192, 195]

[H35]   Note on the number of finite algebras. *J. Combinatorial Theory* **1**
            (1966), 394.   [195]

Havel, V.
[H36]   A remark on the existence of finite graphs (Hungarian). *Časopis Pěst.*
            Mat. **80** (1955), 477–480.   [58]
[H37]   On the completeness-number of a finite graph. *Beitrage zur Graphen-*
            theorie (H. Sachs, H. Voss, and H. Walther, eds.) Teubner, Leipzig,
            1968, pp. 71–74.   [149]

Heawood, P. J.
[H38]   Map colour theorems. *Quart. J. Math.* **24** (1890), 332–338.
                                                             [5, 118, 130, 136]

Hedetniemi, S.
[H39]   On hereditary properties of graphs. *Studia Sci. Math. Hungar.* (to
            appear).   [96, 100]
            See also [CGH1], [CGH2], [HH1], [HHP1], [HHR1], and pp. 140
            and 148.

Hedrlin, Z.
[HP23]   (with A. Pultr), Symmetric relations (undirected graphs) with given
            semigroup. *Monatsh. Math.* **69** (1965), 318–322.   [177]
[HP24]   (with A. Pultr), On rigid undirected graphs. *Canad. J. Math.* **18** (1966),
            1237–1242.   [177]

Heffter, L.
[H40]   Über das Problem der Nachbargebiete. *Ann. Math.* **38** (1891), 477–508.
                                                                            [136]
Hemminger, R. L.
[H41]   On reconstructing a graph. *Proc. Amer. Math. Soc.* **20** (1969), 185–187.
                                                                             [83]

Herz, J. C. See [GHR1].

Heuchenne, C.
[H42]   Sur une certaine correspondance entre graphes. *Bull. Soc. Roy. Sci.*
            Liège **33** (1964), 743–753.   [209]

Hobbs, A. M.
[HG1]   (with J. W. Grossman), Thickness and connectivity in graphs. *J. Res.*
            Nat. Bur. Stand. Sect. B (to appear).   [120]

Hoffman, A. J.
[H43]   On the uniqueness of the triangular association scheme. *Ann. Math.*
            Statist. **31** (1960), 492–497.   [17, 78]
[H44]   On the exceptional case in a characterization of the arcs of complete
            graphs. *IBM J. Res. Develop.* **4** (1960), 487–496.   [78]
[H45]   On the polynomial of a graph. *Amer. Math. Monthly* **70** (1963), 30–36.
                                                                       [159, 210]
[H46]   On the line-graph of the complete bipartite graph. *Ann. Math. Statist.*
            35 (1964), 883–885.   [71, 79]
[HS1]   (with R. R. Singleton), On Moore graphs with diameters 2 and 3. *IBM*
            J. Res. Develop. **4** (1960), 497–504.   [25]
            See also [GH2].

House, L. C.
[H47]   A *k*-critical graph of given density. *Amer. Math. Monthly* **74** (1967), 829–831.   [149]

Izbicki, H.
[I1]   Unendliche Graphen endlichen Grades mit vorgegebenen Eigenschaften. *Monatsh. Math.* **63** (1959), 298–301.   [170]

Jean, M.
[J1]   Edge-similar tournaments. *Recent Progress in Combinatorics* (W. T. Tutte, ed.). Academic Press, New York, 1969.   [211]

Jordan, C.
[J2]   Sur les assemblages de lignes. *J. Reine Angew. Math.* **70** (1869), 185–190.   [36]

            See also p. 35.

Jung, H. A.
[J3]   Zu einem Isomorphiesatz von Whitney für Graphen. *Math. Ann.* **164** (1966), 270–271.   [72]

Kagno, I. N.
[K1]   Linear graphs of degree ≤6 and their groups. *Amer. J. Math.* **68** (1946), 505–520; **69** (1947), 872; **77** (1955), 392.   [176]

Kapoor, S. F. See [CKK1].

Karaganis, J. J.
[K2]   On the cube of a graph. *Canad. Math. Bull.* (to appear).   [69]

Karp, R. M. See [HKT1].

Kasteleyn, P. W.
[K3]   A soluble self-avoiding walk problem. *Physica* **29** (1963), 1329–1337.   [209]

[K4]   Graph theory and crystal physics. Chapter 2 in *Graph Theory and Theoretical Physics* (F. Harary, ed.) Academic Press, London, 1967, pp. 44–110.   [71, 204]

Kaugars, A. See [CKL1] and p. 31.

Kay, D. C.
[KC1]   (with G. Chartrand), A characterization of certain ptolemaic graphs. *Canad. J. Math.* **17** (1965), 342–346.   [24]

Kelly, J. B.
[KK1]   (with L. M. Kelly), Paths and circuits in critical graphs. *Amer. J. Math.* **76** (1954), 786–792.   [128, 142]

Kelly, L. M. See [KK1].

Kelly, P. J.
[K5]   A congruence theorem for trees. *Pacific J. Math.* **7** (1957), 961–968.   [41]

[KM1]   (with D. Merriell), A class of graphs. *Trans. Amer. Math. Soc.* **96** (1960), 488–492.   [25]

Kempe, A. B.
   [K6]   On the geographical problem of four colors. *Amer. J. Math.* **2** (1879), 193–204.   [5]

Kendall, M. G.
   [KS1]   (with B. B. Smith), On the method of paired comparisons. *Biometrika* **31** (1940), 324–345.   [207]

King, C. See pp. 158 and 210.

Kirchhoff, G.
   [K7]   Über die Auflösung der Gleichungen, auf welche man bei der Untersuchung der linearen Verteilung galvanischer Ströme geführt wird. *Ann. Phys. Chem.* **72** (1847), 497–508.   [2, 152, 210]

Klee, V. See [DK1].

Kleinert, M.
   [K8]   Die Dicke des n-dimensionalen Würfel-Graphen. *J. Combinatorial Theory* **3** (1967), 10–15.   [121]

Kleitman, D. See p. 123.

Kodama, Y. See [BHK1], [BHKY1], and [HK1].

König, D.
   [K9]   Graphen und Matrizen. *Mat. Fiz. Lapok* **38** (1931), 116–119.   [53, 96]
 * [K10]   *Theorie der endlichen und unendlichen Graphen.* Leipzig, 1936, Reprinted Chelsea, New York, 1950.   [3, 7, 18, 35, 84, 127, 168]
             See also p. 116.

Kotzig, A.
   [K11]   On certain decompositions of a graph (in Slovakian). *Mat.-Fyz. Časopis* **5** (1955), 144–151.   [55]
             See also pp. 49 and 124.

Krausz, J.   Démonstration nouvelle d'un théorème de Whitney sur les réseaux.
   [K12]
             Mat. Fiz. Lapok **50** (1943), 75–89.   [74]

Kronk, H. V.
   [K13]   Generalization of a theorem of Pósa. *Proc. Amer. Math, Soc.*
                                   [70]
             See also [CKK1] and [CK1].

Kuratowski, K.
   [K14]   Sur le problème des courbes gauches en topologie. *Fund. Math.* **15** (1930), 271–283.   [108]

Landau, H. G.
   [L1]   On dominance relations and the structure of animal societies, III; the condition for a score structure. *Bull. Math. Biophys.* **15** (1955), 143–148.   [207, 211]

Lawes, P. See p. 195.

Lederberg, J. See p. 68.

Lee, T. D.
[LY1]   (with C. N. Yang), Many-body problems in quantum statistical mechanics. *Phys. Rev.* **113** (1959), 1165–1177.   [6]

Lekkerkerker, C. See [BL2].

Lewis, D. See [BL1].

Lewin, K.
* [L2]   *Principles of Topological Psychology*, McGraw-Hill, New York, 1936.   [5]

Lick, D. R. See [CKL1].

Littlewood, D. E.
* [L3]   *The Theory of Group Characters.* Clarendon, Oxford, 1940.   [164]

Lovász, L.
[L4]   On decomposition of graphs. *Studia Sci. Math. Hungar.* **1** (1966), 237–238.   [63]
[L5]   On chromatic number of finite set-systems. *Acta Math. Acad. Sci. Hungar.* **19** (1968), 59–67.   [128]

Lynn, M. S. See [EHL1].

MacLane, S.
[M1]   A structural characterization of planar combinatorial graphs. *Duke Math. J.* **3** (1937), 340–472.   [115]

Manvel, B.
[M2]   Reconstruction of trees. *Canad. J. Math.* (to appear).   [41]
[M3]   Reconstruction of unicyclic graphs. *Proof Techniques in Graph Theory* (F. Harary, ed.) Academic Press, New York, 1969.   [41]
    See also [HM1].

Marczewski, E.
[M4]   Sur deux propriétés des classes d'ensembles. *Fund. Math.* **33** (1945), 303–307.   [19]

May, K. O.
[M5]   The origin of the four-color conjecture. *Isis* **56** (1965), 346–348.   [5]

Mayberry, J. P. See [BM2].

Mayer, J.
[M6]   Le problème des regions voisines sur les surfaces closes orientables. *J. Combinatorial Theory* (to appear).   [119]
    See also p. 120.

McAndrew, M. H.
[M7]   On the product of directed graphs. *Proc. Amer. Math. Soc.* **14** (1963), 600–606.   [21, 210]
[M8]   The polynomial of a directed graph. *Proc. Amer. Math. Soc.* **16** (1965), 303–309.   [177]

Meetham, A. R. See p. 25.

Mendelsohn, N. S. See [DM1], [DM2], and [DM3].

Menger, K.
  [M9]   Zur allgemeinen Kurventheorie. *Fund. Math.* **10** (1927), 96–115.   [47]

Menon, V.
  [M10]   On repeated interchange graphs. *Amer. Math. Monthly* **13** (1966), 986–989.   [71]

Meriwether, R. L. See p. 176.

Merriell, D. See [KM1].

Milgram, A. N. See [GM1].

Miller, D. J.
  [M11]   The categorical product of graphs. *Canad. J. Math.* **20** (1968), 1511–1521.   [25]

Minty, G.
  [M12]   On the axiomatic foundations of the theories of directed linear graphs, electrical networks and network-programming. *J. Math. Mech.* **15** (1966), 485–520.   [41]

Mirsky, L.
  [MP1]   (with H. Perfect), Systems of representatives. *J. Math. Anal. Applic.* **15** (1966), 520–568.   [54, 55]

Moon, J.
  [M13]   On the line-graph of the complete bigraph. *Ann. Math. Statist.* **34** (1963), 664–667.   [79]
  [M14]   An extension of Landau's theorem on tournaments. *Pacific J. Math.* **13** (1963), 1343–1345.   [211]
  [M15]   Various proofs of Cayley's formula for counting trees. Chapter 11 in *A Seminar on Graph Theory* (F. Harary, ed.) Holt, Rinehart and Winston, New York, 1967, pp. 70–78.   [154, 179, 195]
* [M16]   *Topics on Tournaments.* Holt, Rinehart and Winston, New York, 1968.   [195, 198, 211]
  [MM1]   (with L. Moser), On cliques in graphs. *Israel J. Math.* **3** (1965), 23–28.   [25]

          See also [BHM1], [BM1], and p. 83.

Morgenstern, O. See [NM1].

Moser, L. See [HM2] and [MM1].

Motzkin, T. S.
  [MS1]   (with E. G. Straus), Maxima for graphs and a new proof of a theorem of Turán. *Canad. J. Math.* **17** (1965), 533–540.   [19]

Mowshowitz, A.
  [M17]   The group of a graph whose adjacency matrix has all distinct eigenvalues. *Proof Techniques in Graph Theory* (F. Harary, ed.) Academic Press, New York, 1969.   [158]
          See also [HMR1].

Mukhopadhyay, A.
  [M18]   The square root of a graph. *J. Combinatorial Theory* **2** (1967), 290–295.   [24]

Mycielski, J.
[M19]   Sur le coloriage des graphes. *Colloq. Math.* **3** (1955), 161–162.   [128]

Nash–Williams, C. St. J. A.
[N1]   On orientations, connectivity and odd-vertex pairings in finite graphs. *Canad. J. Math.* **12** (1960), 555–567.   [210]
[N2]   Edge-disjoint spanning trees of finite graphs. *J. London Math. Soc.* **36** (1961), 445–450.   [90]
[N3]   Infinite graphs—a survey. *J. Combinatorial Theory* **3** (1967), 286–301.   [16]
           See also [HN1].

von Neumann, J.
* [NM1]   (with O. Morgenstern), *Theory of Games and Economic Behavior.* Princeton University Press, Princeton, 1944.   [202]

Nordhaus, E. A.
[NG1]   (with J. W. Gaddum), On complementary graphs. *Amer. Math. Monthly* **63** (1956), 175–177.   [129]
           See also [BCN1].

Norman, R. Z.
[NR1]   (with M. Rabin), Algorithm for a minimal cover of a graph. *Proc. Amer. Math. Soc.* **10** (1959), 315–319.   [96, 97]
           See also [FNU1], [HN2], [HN3], [HN4], and [HNC1].

Oberschelp, W.
[O1]   Kombinatorische Anzahlbestimmungen in Relationen. *Math. Ann.* **174** (1967), 53–78.   [225]

Ore, O.
[O2]   A problem regarding the tracing of graphs. *Elemente der Math.* **6** (1951), 49–53.   [69]
[O3]   Note on Hamilton circuits. *Amer. Math. Monthly* **67** (1960), 55.   [68]
[O4]   Arc coverings of graphs. *Ann. Mat. Pura Appl.* **55** (1961), 315–322.   [70]
* [O5]   *Theory of Graphs.* Amer. Math. Soc. Colloq. Publ. 38, Providence, 1962.   [21, 62, 71, 128]
[O6]   Hamilton connected graphs. *J. Math. Pures Appl.* **42** (1963), 21–27.   [70]
* [O7]   *The Four Color Problem.* Academic Press, New York, 1967.   [133, 135]
[OS1]   (with G. J. Stemple), Numerical methods in the four color problem. *Recent Progress in Combinatorics* (W. T. Tutte, ed.) Academic Press, New York, 1969.   [5, 132]

Otter, R.
[O8]   The number of trees. *Ann. of Math.* **49** (1948), 583–599.   [188, 195, 196]

Palmer, E. M.
[P1]   Prime line-graphs. *Elem. Math.* (to appear).   [176]
[PR1]   (with R. W. Robinson), The matrix group of two permutation groups. *Bull. Amer. Math. Soc.* **73** (1967), 204–207.   [195]

    See also [HP1], [HP2], [HP3], [HP4], [HP5], [HP6], [HP7], [HP8], [HP9], [HP10], [HP11], [HP12], [HP13], [HP14], [HP15], [HP16], [HP17], [HPR1], and p. 208.

Parthasarathy, K. R.
[P2]   Enumeration of graphs with given partition. *Canad. J. Math.* **20** (1968), 40–47.   [195]

Penney, W. F. See [AFPR1].

Perfect, H. See [MP1].

Petersen, J.
[P3]   Die Theorie der regulären Graphen. *Acta Math.* **15** (1891), 193–220.   [89, 90]

Pippert, R. E. See [BP1].

Plummer, M. D.
[P4]   On minimal blocks. *Trans Amer. Math. Soc.*, **134** (1968), 85–94.   [31]
    See also [BHP1], [BP2], [HP18], [HP19], and pp. 55, 69, and 100.

Pólya, G.
[P5]   Kombinatorische Anzahlbestimmungen für Gruppen, Graphen und chemische Verbindungen. *Acta Math.* **68** (1937), 145–254.   [162, 177, 180, 182, 187, 188, 189]
[P6]   Sur les types des propositions composées. *J. Symb. Logic* **5** (1940), 98–103.   [164, 196]
    See also p. 195.

Pósa, L.
[P7]   A theorem concerning hamilton lines. *Magyar Tud. Akad. Mat. Kutato Int. Kozl.* **7** (1962), 225–226.   [66]
    See also [EGP1].

Powell, M. G. See [WP1].

Prins, G.
[P8]   The automorphism group of a tree. Doctoral dissertation, University of Michigan, 1957.   [195, 231]
    See also [HHP1], [HP20], [HP21], [HP22], and [HPT1].

Pultr, A. See [HP23] and [HP24].

Quintas, L. V. See [GQ1].

Rabin, M. See [NR1].

Rademacher, H. See [SR2].

Radjavi, H. See [BR1].

Rado, R.
[R1]   Note on the transfinite case of Hall's theorem on representatives. *J. London Math. Soc.* **42** (1967), 321–324.   [53]

Ramsey, F. P.
[R2]   On a problem of formal logic. *Proc. London Math. Soc.* **30** (1930), 264–286.   [16]

Read, R. C.
   [R3]   The enumeration of locally restricted graphs. I and II. *J. London Math. Soc.* **34** (1959), 417–436; **35** (1960), 344–351.   [195]
   [R4]   A note on the number of functional digraphs. *Math. Ann.* **143** (1961), 109–110.   [195]
   [R5]   On the number of self-complementary graphs and digraphs. *J. London Math. Soc.* **38** (1963), 99–104.   [192, 195, 197]
   [R6]   An introduction to chromatic polynomials. *J. Combinatorial Theory* **4** (1968), 52–71.   [146, 148]
            See also [HPR1], [HR1], and pp. 158 and 210.

Rédei, L.
   [R7]   Ein kombinatorischer Satz. *Acta Litt. Szeged* **7** (1934), 39–43.   [206]

Redfield, J. H.
   [R8]   The theory of group-reduced distributions. *Amer. J. Math.* **49** (1927), 433–455.   [178]

Reed, M. See [SR1].
Rényi, A. See [ER1].

Richardson, M.
   [R9]   On weakly ordered systems. *Bull. Amer. Math. Soc.* **52** (1946), 113–116.
                                                                                [202]
Riddell, R. J.
   [RU1]   (with G. E. Uhlenbeck), On the theory of the virial development of the equation of state of monoatomic gases. *J. Chem. Physics* **21** (1953), 2056–2064.   [195]

Ringel, G.
* [R10]   *Färbungsprobleme auf Flachen und Graphen.* Deutscher Verlag der Wissenschaften, Berlin, 1962.   [119]
   [R11]   Selbstkomplementäre Graphen. *Arch. Math.* **14** (1963), 354–358   [24]
   [R12]   Das Geschlecht des vollständiger paaren Graphen. *Abh. Math. Sem. Univ. Hamburg* **28** (1965), 139–150.   [119]
   [R13]   Über drei kombinatorische Probleme am *n*-dimensionalen Würfel und Würfelgitter. *Abh. Math. Sem. Univ. Hamburg* **20** (1955), 10–19.   [119]
   [RY1]   (with J. W. T. Youngs), Solution of the Heawood map-coloring problem. *Proc. Nat. Acad. Sci. USA* **60** (1968), 438–445.   [119]
   [RY2]   (with J. W. T. Youngs), Remarks on the Heawood conjecture. *Proof Techniques in Graph Theory* (F. Harary, ed.) Academic Press, New York, 1969.   [119]
            See also p. 125.

Riordan, J.
   [R14]   The number of labelled colored and chromatic trees. *Acta Math.* **97** (1957), 211–225.   [190, 195]
* [R15]   *An Introduction to Combinatorial Analysis.* Wiley, New York, 1958.
                                                                          [145, 213, 231]
   [R16]   The enumeration of trees by height and diameter. *IBM J. Res. Develop.* **4** (1960), 473–478.   [195]
            See also [AFPR1], [CR1], [GR1], and [HMR1].

Robbins, H. E.
  [R17]   A theorem on graphs with an application to a problem of traffic control.
            Amer. Math. Monthly **46** (1939), 281–283.   [210]
            See also [CR1].

Robertson, N.
  [R18]   The smallest graph of girth 5 and valency 4. *Bull. Amer. Math. Soc.*
            70 (1964), 824–825.   [177]
            See also p. 74.

Robinson, R. W.
  [R19]   Enumeration of colored graphs. *J. Combinatorial Theory* **4** (1968),
            181–190.   [195]
            See also [HHR1], [PR1], and p. 70.

van Rooij, A.
  [RW1]   (with H. Wilf), The interchange graphs of a finite graph. *Acta Math.
            Acad. Sci. Hungar.* **16** (1965), 263–269.   [74]

Ross, I. C. See [HR2].

Rossi, P. See [GHR1].

Rota, G.-C.
  [R20]   On the foundations of combinatorial theory, I: Theory of Möbius
            functions. *Z. Wahrscheinlichkeitstheorie und Verw. Gebiete* **2** (1964),
            340–368.   [147]

Ryser, H. J.
  [R21]   Matrices of zeros and ones. *Bull. Amer. Math. Soc.* **66** (1960), 442–464.
                                                                             [209]

Saaty, T. See p. 123.

Sabidussi, G.
  [S1]   Loewy-groupoids related to linear graphs. *Amer. J. Math.* **76** (1954),
            447–487.   [162]
  [S2]   Graphs with given group and given graph-theoretical properties.
            Canad. J. Math. **9** (1957), 515–525.   [170, 176]
  [S3]   On the minimum order of graphs with given automorphism group.
            Monatsh. Math. **63** (1959), 124–127.   [176]
  [S4]   The composition of graphs. *Duke Math. J.* **26** (1959), 693–696.   [166]
  [S5]   Graph multiplication. *Math. Z.* **72** (1960), 446–457.   [21, 166]
  [S6]   The lexicographic product of graphs. *Duke Math. J.* **28** (1961), 573–578.
                                                                             [21]
  [S7]   Graph derivatives. *Math. Z.* **76** (1961), 385–401.   [71]

Sachs, H.
  [S8]   Über selbstkomplementäre Graphen. *Publ. Math. Debrecen* **9** (1962),
            270–288.   [24]
  [S9]   Regular graphs with given girth and restricted circuits. *J. London Math.
            Soc.* **38** (1963), 423–429.   [93]
            See also [FS1].

Schuster, S.   See [DS1].

Sedláček, J.
[S10]   Some properties of interchange graphs. *Theory of Graphs and its Applications* (M. Fiedler, ed.) Prague, 1962; Reprinted, Academic Press, New York, 1962, pp. 145–150.   [124]

Senior, J. K.
[S11]   Partitions and their representative graphs. *Amer. J. Math.* **73** (1951), 663–689.   [63]

Seshu, S.
* [SR1]   (with M. Reed), *Linear Graphs and Electrical Networks*. Addison-Wesley, Reading, 1961.   [71]

Shannon, C. E. See [EFS1].

Shrikhande, S. S.
[S12]   On a characterization of the triangular association scheme. *Ann. Math. Statist.* **30** (1959), 39–47.   [79]

Singleton, R. R.
[S13]   There is no irregular Moore graph. *Amer. Math. Monthly* **75** (1968), 42–43.   [24]
         See also [HS1].

Slepian, D.
[S14]   On the number of symmetry types of Boolean functions of n variables. *Canad. J. Math.* **5** (1953), 185–193.   [195]

Smith, B. B. See [KS1].

Smith, C. A. B.
[ST1]   (with W. T. Tutte), On unicursal paths in a network of degree 4. *Amer. Math. Monthly* **48** (1941), 233–237.   [195, 204]
         See [BSST1] and pp. 68 and 123.

Stein, S. K.
[S15]   Convex maps. *Proc. Amer. Math. Soc.* **2** (1951), 464–466.   [106]

Steinitz, E.
* [SR2]   (with H. Rademacher), *Vorlesungen über die Theorie der Polyeder*. Springer, Berlin, 1934.   [106]

Stemple, G. J. See [OS1].

Stewart, M. J. See [CS1].

Stone, A. H. See [BSST1] and p. 123.

Straus, E. G. See [MS1].

Sylvester, J. J. See pp. 1, 3, and 35.

Szekeres, G.
[SW1]   (with H. S. Wilf), An inequality for the chromatic number of a graph. *J. Combinatorial Theory* **4** (1968), 1–3.   [127]
         See also [ES1].

Szele, T.
[S16]   Kombinatorische Untersuchungen über den gerichteten vollständigen graphen. *Mat. Fiz. Lapok* **50** (1943), 223–256. (In Hungarian.)   [206]

Tait, P. G.
[T1]   Remarks on the colouring of maps. *Proc. Royal Soc. Edinburgh* **10** (1880), 729. [68]

Tang, D. T.
[T2]   Bi-path networks and multicommodity flows. *IEEE Trans. Circuit Theory* **11** (1964), 468–474. [124]

Teh, H, H,
[TY1]   (with H. D. Yap), Some construction problems of homogeneous graphs. *Bull. Math. Soc. Nanyang Univ.* (1964), 164–196. [21]

Trauth, C. A., Jr. See [HT1] and [HT2].

Turán, P.
[T3]   Eine Extremalaufgabe aus der Graphentheorie. *Mat. Fiz. Lapok* **48** (1941), 436–452. (In Hungarian.) [17, 18, 25]

Turner, J.
[T4]   Point-symmetric graphs with a prime number of points. *J. Combinatorial Theory* **3** (1967), 136–145. [176]
[T5]   A bibliography of graph theory. *Proof. Techniques in Graph Theory* (F Harary, ed.) Academic Press, New York, 1969. [235]

Tutte, W. T.
[T6]   On Hamilton circuits. *J. London Math. Soc.* **21** (1946), 98–101. [123]
[T7]   The factorizations of linear graphs. *J. London Math. Soc.* **22** (1947), 107–111. [85, 92]
[T8]   A family of cubical graphs. *Proc. Cambridge Philos. Soc.* **43** (1947), 459–474. [175]
[T9]   The dissection of equilateral triangles into equilateral triangles. *Proc. Cambridge Philos. Soc.* **44** (1948), 463–482. [203]
[T10]   The factors of graphs. *Canad. J. Math.* **4** (1952), 314. [88]
[T11]   A short proof of the factor theorem for finite graphs. *Canad. J. Math.* **6** (1954), 347–352. [88]
$[T11\frac{1}{2}]$ A theorem on planar graphs, *Trans. Amer. Math. Soc.*, **82** (1956) 99–116. [123]
[T12]   An algorithm for determining whether a given binary matroid is graphic. *Proc. Amer. Math. Soc.* **11** (1960), 905–917. [157]
[T13]   A theory of 3-connected graphs. *Indag. Math.* **23** (1961), 441–455. [46]
[T14]   A census of planar triangulations. *Canad. J. Math.* **14** (1962), 21–38. [195]
[T15]   A new branch of enumerative graph theory. *Bull. Amer. Math. Soc.* **68** (1962), 500–504. [193]
[T16]   On the non-biplanar character of the complete 9-graph. *Canad. Math. Bull.* **6** (1963), 319–330. [108]
[T17]   How to draw a graph. *Proc. London Math. Soc.* **13** (1963), 743–767. [112]
[T18]   The number of planted plane trees with a given partition. *Amer. Math. Monthly* **71** (1964), 272–277. [195]
[T19]   Lectures on matroids. *J. Res. Nat. Bur. Stand.* Sect. B **69** (1965), 1–47. [41, 156]

* [T20]   *The Connectivity of Graphs.* Toronto Univ. Press, Toronto, 1967.
[173, 174, 177]
See also [BSST1], [HKT1], [HPT1], [HT3], [HT4], [ST1], and pp. 123 and 128.

Uhlenbeck, G. E.
[U1]   Successive approximation methods in classical statistical mechanics. *Physica* **26** (1960), 17–27.   [6]
See also [FNU1], [HU1], and [RU1].

Ulam, S. M.
* [U2]   *A Collection of Mathematical Problems.* Wiley (Interscience), New York, 1960.   [12]

Vajda, S.
* [V1]   *Mathematical Programming.* Addison-Wesley, Reading, 1961.   [7]

Varga, R. S.
* [V2]   *Matrix Iterative Analysis.* Prentice Hall, Englewood Cliffs, 1962   [6]

Veblen, O.
[V3]   *Analysis Situs.* Amer. Math. Soc. Colloq. Publ. Vol. 5, Cambridge, 1922. Second edition, New York, 1931.   [7]

Vizing, V. G.
[V4]   On an estimate of the chromatic class of a p-graph (Russian) *Diskret. Analiz.* **3** (1964), 25–30.   [133]
[V5]   On the number of edges in a graph with given radius (Russian) *Dokl. Akad. Nauk. SSSR* **173** (1967), 1245–1246.   [42]

Vollmerhaus, H.
[V6]   Über die Einbettung von Graphen in zweidimensionale orientierbare Mannigfaltigkeiten kleinsten Geschlechts. *Beitrage zür Graphentheorie* (H. Sachs, H. Voss, and H. Walther, eds.) Teubner, Leipzig, 1968, pp. 163–168.   [117]

Wagner, K.
[W1]   Bemerkungen zum Vierfarbenproblem. *Jber. Deutsch. Math.-Verein.* **46** (1936), 26–32.   [106]
[W2]   Über eine Eigenschaft der ebenen Komplexe. *Math. Ann.* **114** (1937), 570–590.   [113]
[W3]   Bemerkungen zu Hadwigers Vermutung, *Math. Ann.*, **141** (1960), 433–451.   [135]

Walther, H.
[W4]   On intersections of paths in a graph. *J. Combinatorial Theory* (to appear).
[24]

Watkins, M. E.
[W5]   A lower bound for the number of vertices of a graph. *Amer. Math. Monthly* **74** (1967), 297.   [56]

Weichsel, P. M.
[W6]   The Kronecker product of graphs. *Proc. Amer. Math. Soc.* **13** (1963), 47–52.   [21]

Weinberg, L.
  [W7]   Number of trees in a graph. *Proc. IRE* **46** (1958), 1954–1955.   [196]
  [W8]   On the maximum order of the automorphism group of a planar triply
            connected graph. *SIAM J.* **14** (1966), 729–738.   [177]

Welsh, D. J. A.
  [W9]   Euler and bipartite matroids. *J. Combinatorial Theory* (to appear).
                                                                                      [159]
  [WP1]   (with M. B. Powell), An upper bound for the chromatic number of a
            graph and its application to timetabling problems. *Computer J.* **10**
            (1967), 85–87.   [148]
            See also p. 149.

Whitney, H.
  [W10]   The coloring of graphs. *Ann. Math. (2)* **33** (1932), 688–718.   [147]
  [W11]   Congruent graphs and the connectivity of graphs. *Amer. J. Math.* **54**
            (1932), 150–168.   [43, 48, 71, 72, 177]
  [W12]   Non-separable and planar graphs. *Trans Amer. Math. Soc.* **34** (1932),
            339–362.   [104, 113, 123]
  [W13]   A set of topological invariants for graphs. *Amer. J. Math.* **55** (1933),
            231–235.   [105]
  [W14]   Planar graphs. *Fund. Math.* **21** (1933), 73–84.   [113]
  [W15]   On the abstract properties of linear dependence. *Amer. J. Math.* **57**
            (1935), 509–533.   [40]
  [W16]   A theorem on graphs. *Annals Math.* **32** (1931), 378–390.   [133]

Wielandt, H.
  [W17]   Unzerlegbare, nichtnegative Matrizen. *Math. Z.* **52** (1950), 642–648.
                                                                                      [210]

Wilcox, G. See [HW1].

Wilf, H. S. See [RW1] and [SW1].

Yackel, J. See [GY1].

Yang, C. N. See [LY1].

Yap, H. D. See [TY1].

Youngs, J. W. T.
  [Y1]   The Heawood map colouring conjecture. Chapter 12 in *Graph Theory
            and Theoretical Physics* (F. Harary, ed.) Academic Press, London, 1967,
            pp. 313–354.   [119]
            See also [BHKY1], [RY1], and [RY2].

Zykov, A. A.
  [Z1]   On some properties of linear complexes. (Russian) *Mat. Sbornik* **24**
            (1949), 163–188. *Amer. Math. Soc. Translation* N. 79, 1952.
                                                                           [21, 128, 146]

INDEX OF SYMBOLS

INDEX OF SYMBOLS

The Greeks had a word for it ...

Z. Akins

Most of the letters in the Roman and Greek alphabets have been used as symbols in this book. Those symbols which occur most often are listed here, separated into three categories: Roman letters, Greek letters, and operations on graphs and groups.

A	adjacency matrix 150, 151		Q_n	n-cube 23
A_p	alternating group 165		$S(G)$	subdivision graph of G 81
B	incidence matrix 152		S_p	symmetric group 165
$B(G)$	block graph of G 29		$S_p^{(2)}$	pair group 185
C	cycle matrix 154		$S_p^{[2]}$	reduced ordered pair group 186
C^*	cocycle matrix 155			
C_n	cycle of length n 13		T	tree 32
C_p	cyclic group 165		T	tournament 205
$C(G)$	cutpoint graph of G 30		T^*	cotree of T 39
D	digraph 198		$T(G)$	total graph of G 84
D^*	condensation of D 200		V	set of points of G 9
D'	converse of D 200		W_n	wheel 46
D_p	dihedral group 165		X	set of lines 9
E_p	identity group 165		$Z(A)$	cycle index of A 181
G	graph 9			
$G - u$	removal of a point 11		$bc(G)$	block-cutpoint tree of G 36
$G - x$	removal of a line 11		$c(G)$	circumference 13
$G + x$	addition of a line 11		d_i	degree of v_i 14
G^2	square of G 14		$d(G)$	diameter 14
G^*	dual of G 113		$d(u, v)$	distance 14, 199
K_p	complete graph 16		$e(v)$	eccentricity 35
$K_{m,n}$	complete bigraph 17		$g(G)$	girth 13
$L(D)$	line digraph of D 209		$id(v)$	indegree 198
$L(G)$	line graph of G 71		$j_k(\alpha)$	number of k-cycles 181
P_n	path 13		$k(G)$	number of components 40

$m(G)$	cycle rank	39
$m^*(G)$	cocycle rank	39
$od(v)$	outdegree	198
p	number of points	9
(p, q)	p points, q lines	9
q	number of lines	9
$r(G)$	radius	35
u, v, w	points	9
x, y, z	lines	9

α_0 point covering number   94

α_1 line covering number   94

β_0 point independence number   95

β_1 line independence number   95

γ genus   117

$\Gamma(G)$ group of G 161

$\Gamma_1(G)$ line group of G 161

δ minimum degree   14

Δ maximum degree   14

θ thickness   120

κ connectivity   43

$\kappa(u, v)$ local connectivity   49

λ line-connectivity   43

ν crossing number   122

ξ coarseness   121

$\Pi(G)$ partition of a graph   57

$\Upsilon(G)$ arboricity   90

χ chromatic number   127

χ' line-chromatic number   133

ψ achromatic number   144

ω intersection number   19

$\Omega(F)$ intersection graph   19

$\langle S \rangle$ induced subgraph   11

$G_1 \cup G_2$ union of graphs   21

$G_1 + G_2$ join of graphs   21

$A + B$ sum of groups   163

$G_1 \times G_2$ product of graphs   21

$A \times B$ product of groups   163

$G_1[G_2]$ composition of graphs   22

$A[B]$ composition of groups   164

$G_1 \wedge G_2$ conjunction of graphs   25

B^A power group   164

$G_1 \circ G_2$ corona of graphs   167

INDEX OF DEFINITIONS

INDEX OF DEFINITIONS

In words, as fashions, the same rule will hold,
Alike fantastic if too new or old;
Be not the first by whom the new are tried,
Nor yet the last to lay the old aside.
A. POPE, *Essay on Criticism*

achromatic number, 144
acyclic, digraph, 200
  graph, 32
addition of a line, 11
adjacency matrix, of a digraph, 151, 202
  of a graph, 150
adjacent lines, 9
adjacent points, in a digraph, 198
  in a graph, 9
animal, 194
arbitrarily traversable graph, 69
arboricity, 90
arc, 10
automorphism, 161

1-basis, 202
bigraph, 17
  complete, 17
bipartite graph, 17
block, 26
block graph, 29
block-cutpoint graph, 36
  tree, 37
boundary, 37
branch, 35
bridge, 26

n-cage, 174
center, 35
central point, 35

centroid, 36
centroid point, 36
0-chain, 37
1-chain, 37
chord, 38
n-chromatic graph, 127
chromatic number, of a graph, 127
  of a manifold, 135
n-chromatic number, 149
chromatic polynomial, 146
circuits, 40, 41
circumference, 13
clique, 20
clique graph, 20
coarseness, 121
coboundary, 38
cocircuit, 41
cocycle, 38
cocycle basis, 38
  matrix, 155
  rank, 39
  space, 38
color class, 126
color-graph, 168
n-colorable graph, 127
  map, 131
coloring, 126
  complete, 143
  of a graph from t colors, 145

ABCDEFGH798765432